Seizing the Means of Reproduction

EXPERIMENTAL FUTURES

Technological Lives, Scientific Arts, Anthropological Voices

A series edited by Michael M. J. Fischer and Joseph Dumit

Seizing the Means of Reproduction

Entanglements of Feminism, Health, and Technoscience

MICHELLE MURPHY

Duke University Press *Durham and London* 2012

© 2012 Duke University Press

All rights reserved

Printed in the United States of
America on acid-free paper ∞

Typeset in Chaparral Pro by
Tseng Information Systems, Inc.

Library of Congress Cataloging-in-
Publication Data appear on the last
printed page of this book.

CONTENTS

Acknowledgments vii

Introduction. Feminism in/as Biopolitics 1

1 Assembling Protocol Feminism 25

2 Immodest Witnessing, Affective Economies,
and Objectivity 68

3 Pap Smears, Cervical Cancer, and Scales 102

4 Traveling Technology and a Device for Not
Performing Abortions 150

Conclusion. Living the Contradiction 177

Notes 183

Bibliography 219

Index 247

ACKNOWLEDGMENTS

When I started this project ten years ago I imagined I would be writing a history of the United States women's health movement. Yet once I started following feminist techniques, letting their travels lead my research in and out of feminisms, I began posing new questions about the histories of appropriation and transformation through time, place, and politics. In this task, I have many people to thank.

I am deeply grateful to the participants in the feminist self help movement who generously shared time and information with me: Carol Downer, Suzanne Gage, Eileen Schnitger, Shawn Heckert, Monika Franznick, Nancy Boothe, Paula Hammock, Peggy Antrobus, and Andaiye. In addition, I would like to acknowledge the enthusiastic assistance of Lorraine Rothman and Dido Hasper, two important figures of the California feminist health movement who both died during the writing of this book. I am also thankful to the Boston Women's Health Collective Library and the Schlesinger Library for their assistance in collecting materials.

This project was generously supported by the Social Science and Humanities Research Council of Canada. Support for the project also came from the Max Planck Institute for the History of Science, the Peter Wall Institute at the University of British Columbia, the Clayman Center for Gender Research at Stanford, and the Jackman Humanities Institute at the University of Toronto. I wish to thank Courtney Berger at Duke University Press for guiding the improvement of the manuscript.

Over the many years of working on this book, I incurred debts to many colleagues and friends for their generous intellectual support. At the MPIWG, my sincere thanks to Lorraine Daston, Abigail Lustig, Sabine Höhler, Chris Kelty, Maureen McNeill, Hannah Landecker, and the Berlin Feminist Science Studies Workgroup. The arguments in this book are part of larger intellectual debates in the field of science and technology studies, and hence I am indebted to many colleagues in this field for their critical intellectual engagement: Gregg Mitman, Joe Dumit, Warwick Anderson, Lochlann Jain, Cori Hayden, Jackie Orr, Joe Masco, Steven Epstein, Evelynn Hammonds, Samer Alatout, Vincanne Adams, and the members of the Oxidate workgroup. I would especially like to acknowledge Kavita Philip's influence: her incisive feedback and inspiring exchanges have had a deep influence. Since I was a graduate student, Adele Clarke has acted as a crucial mentor, despite the fact that I never quite managed to become located in the San Francisco Bay area for any length of time. Her imprint on this book is large.

The University of Toronto has provided me with a wonderful environment in which to stretch my thinking about transnational feminist analytics. In particular, I would like to acknowledge the roles that Alissa Trotz, Linzi Manicom, Shahrzad Mojab, Ashwini Tambe, Elspeth Brown, Mariana Valverde, and Ritu Birla have had on my politics and thinking. I have had the privilege of working with wonderful graduate students at the University of Toronto who have helped me to refine my approach to the history of feminism and its entanglements with race and science, with Carla Hustak, Rachel Berger, Michael Pettit, and Brian Beaton playing particularly important roles in shaping this book. My collaborators in the Toronto-based Technoscience Salon have been crucial interlocutors. I am particularly grateful to Natasha Myers for her support, both intellectual and social, in our mutual efforts to form a charged technoscience studies community accountable to contemporary politics and thought. I would also like to thank Andrea Adams, Jason Brown, Kathryn Scharf, and Sean Fitzpatrick for their encouragement and friendship

Finally, my greatest debt goes to my lovely family. Claudette and Ted Murphy and Ellen and Richard Price have helped this project through their generosity through thick and thin. On the day I sent my final revisions to the press, my grandmother Loretta Yuill died, and so I honor her here. I have been buoyed by Mika and Maceo Mercey, who remind me what life is all about. I am immeasurably grateful to Matthew Price, who has read every word of this book many times over. His judicious editing and keen eye for writing have improved this work tremendously. He has propped me up in incalculable ways during the years it has taken me to complete this work and continues to amaze me with his generosity. Thank you.

Feminism in/as Biopolitics

Sex changed in the second half of the twentieth century. With the aid of synthetic hormones, immortal tissue cultures, and delicate pipettes the very biological processes of human fertility, and even the sexual form of the body as male and female, became profoundly manipulable. Labs and clinics were vital spaces to this transformation of sex, but so too were state departments of finance and aid agencies, as well as supranational organizations such as the World Bank. Large-scale national and transnational schemes encouraged the technological limiting of births, distributing birth control pills, IUDs, and surgical sterilization to millions, helping to alter the fertility of entire populations for the sake of a greater economic good. The alterability of reproduction in its aggregate form—as "population"—became a shifting planetary problem amenable to technical, state, and market solutions. Sex's changeability expanded further, beyond humans, to intensify in the animal and plant kingdoms as agribusiness mutated seeds into patentable commodities, and livestock was bred with artificial insemination and embryo transfer. This rapidly emerging technical ability to alter human and nonhuman reproduction, stretching from molecular to transnational economic scales, was accompanied by new problems and promises for the politicization of life—not just should, but *how* could reproduction be transformed?

Feminists in California during the 1970s answered this promise by politicizing the details of biomedical practice. They appropriated, revised, and invented reproductive health care techniques: making photographic diaries of cervical variation, crafting politicized health manuals, examining menstruation with a microscope, building an abortion device with a canning jar and aquarium tubing, forming artificial insemination groups, or turning a living room into a health clinic. Each of these tactics heralded the alterability of sex. In manifestos, speeches, posters, T-shirts, court cases, and protest signs, feminists of diverse aims declared the imperative to *seize the means of reproduction*.

In the 1970s, such activities reimagined the promissory orientation of feminism, turning the practical project of managing reproduction into a "necessary," though not sole, political goal of feminist politics. Participants in the particular strand of feminism called the feminist self help movement, who founded some of the first feminist health centers in the United States, sought to concretely rearrange the material, technical, and social conditions by which the responsibility for governing sex could be bound to women as individuals—not the state, experts, or market forces. In these ways, women were "responsibilized" in sex's alterability.[1] In California during the 1970s, as biotechnology was beginning to transform the cells and genetics of sex into opportunities for venture capital, and as United States foreign policy embraced the "population bomb" as a Cold War danger, many feminists also sought to govern the alterability of reproduction through varied low-tech, do-it-yourself, and local interventions.

This book is about a set of feminist projects in the 1970s, 1980s, and beyond that mobilized, politicized, and experimented with technoscience as a means to fashion the "control" of reproduction as a practical and pivotal feature of feminist politics, a set of projects typically grouped together and celebrated, without much historical specificity, as the "women's health movement."[2] This book argues that feminist efforts to remake the terms of medical care and research were at once a critical diagnosis of late twentieth-century technoscience and symptomatic of broader historical shifts. The book attempts to capture the multiple, friction-filled, and yet productive relationships between feminism, as a kind of counter-conduct, and late twentieth-century technoscience, attending to how both feminism and technoscience were each entangled and variegated formations.[3] In so doing, it portrays these entanglements through the histories of how particular techniques moved and were remade between places and times.

A second premise of the book is the converse of the first: that is, technoscience and the politicization of living-being have been defining features of late twentieth-century feminisms. In the late twentieth century many feminisms crafted their politics out of technoscience and bodies, rather than, say, labor or citizenship. Such feminisms participated in making legible the ways technologies and techniques were imbued with politics. Instead of an overarching history of the women's health movement, this book charts how specific feminisms were diversely animated by and entangled with histories of medicine, subjectivities, race, governance, and capitalism in the late twentieth century. It tells stories about the conditions of possibility for feminism as a kind of technoscientific counter-conduct, taking as its starting point the specificity of California in the 1970s within a nation riven by the work of racialized and economic difference in the shadow of American imperial ambitions.

Feminist health projects can be historicized with the same critical analytics used to consider any other instance of technoscience or governance. For example, feminist health practices of the late twentieth century often shared with emerging neoliberal practices an ethic of fashioning inexpensive and individualized interventions into health problems. While high-tech reproductive technologies, such as genetics and cloning, have garnered more attention from scholars as a route to understanding the tangle of life, politics, and capitalism, the less glamorous and simpler technologies examined in this book have vitally touched a vastly greater number of people and have provided crucial sites for the emergence of neoliberal governmentalities, for the industrialization of medicine, and for the entanglement of sexed and raced living-being with capitalism.

In 1970, when this book begins its story, transforming sexed living-being was a shifting technical problem caught in a range of political projects. Not only was life alterable in new ways, but the practices that made up the life and human sciences were experimental systems themselves undergoing reassembly.[4] In the broader context of the Cold War, this experimental ethos expanded to incorporate large regions of the world as testing sites for "development" projects that often saw reproduction and health as pivotal problems in need of intervention. Postcolonial circuits of mobile technical practices rendered each new site of development an opportunity for more data extraction and protocol revision. By the 1970s Cold War United States–sponsored economic development brought the surveillance and reduction of regional birthrates into the heart of its project, positing fertility reduction as a necessary condition for the successful staging of industrialization, green revolutions, free market governance, and other assorted projects of modernization. Such projects were rife with imperatives to simultaneously alter bodies and economies, improving the health of the one, raising the GDP of the other. To this end, new professions of experts proliferated to calibrate and alter the "global facts of life," surveying the world down to the person, counting who had and had not joined the project of regulating fertility in the name of modernity.[5] It is within this larger set of historical shifts (among others) that I want to place the local history of feminist reproductive health projects in California, as entangled with and not independent from these transnational histories.

In the 1970s, then, no matter where in the world you lived, a family-planning clinic may well have opened near you. You may have been enjoined through billboards or posters to plan your fertility with the help of new commodities now distributed globally for just that purpose. Or perhaps, a social worker knocked at your door. You may have been offered the insertion of a Lippes Loop, or the convenience of Blue Lady brand pills, or the popular procedure of surgical sterilization. Sex's alterability presented so many directions to choose from. But maybe the clinic did not just enjoin, but cajoled, targeted, or even coerced. Maybe sterilization was violently enforced. Maybe the state, or the clinic, or even the doctor evaluated your body in terms of racial fitness or economic vitality. Or perhaps the mandate of the clinic took no interest in you and your kind. Quite possibly the choices were few: the pills had side effects, the device was painful, the surgery botched. The sociotechnical experiments in and outside of clinics created whole new ways of being an experimental subject.

Millions of women encountered this experimental mode of technoscience in deeply stratified ways. Moreover, the technical means of altering reproduction was not confined to experts; it was also performed by women themselves. While the contradictions of reproductive politics could foster the coercive management of the racialized poor, so too did it enjoin the participation and enthusiasm of consumers and users, and even projects of "liberation" from the tyrannies of sexed-being. The propagation of cheap disposable medical commodities, the new ease of circulating photocopied information, the simplicity of health procedures now delegated to technicians all helped to make it more possible than ever to learn how to play with technoscience, to turn it into your own project. What is commonly called "the women's health movement" was just such an example of doing technoscience differently, of technoscience as a popular counter-conduct at the nexus of oppression and enjoinment.

Not only feminists, but radicals of many stripes, took up critical and experimental engagements with technoscience in the 1970s. In the United States, the free clinic, the community clinic, and the women's health center became nonprofit alternatives to what was then being newly named the "medical industrial complex."[6] The Black Panthers offered sickle cell screening and health services alongside warm food and child care at community centers.[7] Popular forms of environmental science emerged as ordinary people sought to map and document the toxic conditions in which they lived, played, and work.[8] Sustainable farming experiments countered global agribusiness. The present-day academic field of science and technology studies blossomed in the 1970s, and in many ways sits in a genealogic relation to these political projects.

As living processes became newly open to alteration in the final third of the twentieth century, the possible positions taken by feminists proliferated. In this moment—particularly the 1970s and 1980s—many feminists in the United States and elsewhere believed they could go beyond transforming the practices of health care. They could potentially "seize the means of reproduction," that is, technically manipulate their very embodied relationship to sexed living-being itself. Not only was a feminist critique of science and technology declared imperative; technical practices themselves were possibly the means, the necessary tools, of feminism.

For example, in the United States, just such an explicit call to seize the means of reproduction was made by Shulamith Firestone, a leading figure in the rarified, largely white, radical feminist circles of the urban Northeast: Chicago, New York, Boston. Like Marx and Engels, who had theorized that the proletariat needed to seize the means of production in order to smash the bonds of capitalist relations, Firestone argued that women needed to seize the means of reproduction in order to sever the chains of a patriarchy that fundamentally depended on the uneven material distribution of mammalian biological reproductive labor into male and female bodies. In Firestone's bestselling book, *The Dialectic of Sex* (1970), altering reproductive relations took primacy over any reordering of sexed-labor division in capitalism. Just as the communist revolution would only be complete once all class differences were destroyed, so too would the hoped-for feminist revolution only succeed once all sex categories were unnecessary, a future that could be accomplished if reproductive processes themselves were materially redistributed with the aid of technoscience—or potentially even removed from the body altogether.[9] Though Firestone's revolutionary manifesto might seem dated or absurd, it appealed widely to an emergent feminist moment that took the flesh and well-being of sexed bodies as necessarily alterable and political, thereby hinging questions of freedom and oppression to those of technoscience and life.

While Firestone's suggestion that embodied reproduction be materially abolished was indeed extreme in the 1970s, over the 1980s the technical problem of altering reproduction and "women's health" became central concerns of international development, governance, and health care, with feminists occupying the many professions that made up these domains. As the NGO became the dominant organizational unit for feminism, with outposts in every region of the globe, so too did the women's health NGO become the most common kind of feminist NGO.[10] Twenty years after Firestone's polemic, feminists with advanced degrees served as experts at the UN or the World Bank on matters of reproductive health. This NGO-ization and professionalization of feminism suggests that we need new ways of telling the history of feminism beyond categories of kinds of feminist ideologies. By the end of the 1980s, feminisms (as a diverse tradition of counter-conduct) and broader institutional formations (such as development and family planning) were profoundly caught up in and animating each other through layered and looping histories of mutual appropriation. Thus, this book seeks to experiment with another way of telling the history of feminism, through stories of how feminisms (in the plural) were made through entanglements with broader historical formations, in this case technoscience, public health, neoliberalism, racial formations, and family planning.

Amidst these entanglements, "reproduction," was itself not an obvious phenomenon. Reproduction was not a biological thing with clear bounds, but a multifaceted and distributed effect in time and space, a problem both material and political to which questions of state, race, freedom, individuality, and economic prosperity were bound in ways that connected the micrological with the transnational via embodiment. Just as reproduction was a multidimensional and unevenly distributed problem, the question of how feminists, (as activists, NGO workers, or expert professionals, as well as embodied sexed subjects) sought to understand and intervene in reproduction formed its own disunified field of action. What to change about reproduction? Where did reproduction begin and end? And what about biomedicine needed altering? Who did the changing? Should reproduction be unhitched from economy? How was health tied to liberation? How to craft alterity out of life's alterability? The problem of women's health, though prominent in the period from 1970 to the end of the century, was not at all self-evident or univalent.

While some feminists founded women's health clinics, and some worked within professional medicine or within state agencies to change practices or policies, others critiqued the links between some feminisms and racialized state projects to curb poverty through population control, rejecting biomedical and demographic framings of the problem of reproduction as population. In 1989, in rural Bangladesh, in the town of Comilla (an iconic site of international development projects) feminist activists, professionals, and intellectuals, primarily from South Asia and Europe, wrote a declaration critiquing the ways Cold War experiments had so thoroughly tied reproduction together with capitalism and technoscience. Moreover, they argued that this knot was facilitated by the ways liberal feminists had consolidated around a vision of a universalized female ethical subject who just needed her reproductive rights to do right. The Declaration of Comilla was important for the way it drew distinctions between different feminist projects, as well as situated the politics of reproduction within the "engineering and industrialization of the life processes" more broadly.[11] The forum's most forceful theorizer was also its main organizer, Farida Akhter, who insisted that the "relations of reproduction" could not be severed from the relations of production.[12] For the Declaration of Comilla, reproduction stretched beyond bodies to implicate the multiple domains of industrialism and its environmental effects, family formations, agriculture, and the ownership of biodiversity, thereby necessitating a sweeping critique of technoscience, colonialism, and capitalism. Their vision of an expansive reproductive politics was not remediable by the free choices of an individualized ethical subject. Yet, the declaration did not conclude with a blanket rejection of technoscience (as some other 1980s feminists promulgated); instead, it kept open a promissory future of technoscience done differently, a hope for a possible technoscience that was "error friendly and contributed to the preserving of biological, cultural and social diversity of all living beings."[13] Between Firestone and the Comilla Declaration grew a panoply of ways feminists, technoscience, capitalism, and reproduction could be tied together.

A particular set of tactics for "seizing the means of reproduction" offers the primary entry point of this book: the feminist self help movement of Southern California in the 1970s and 1980s as assembled by a group of women, largely white, who sought to craft feminist techniques of health care and research. While it is tempting to judge past feminisms for their errors or celebrate them for their successes—a kind of historiography that the historian Brian Beaton aptly names "slap or clap"—this book's consideration of feminist health projects seeks to step back to *historicize* their social and technoscientific practices as they were assembled, animated, and entangled within larger biopolitical conjunctures of the twentieth century.[14] Moreover, these feminist health projects provide a lens into the ways large-scale changes in technoscience, governance, and capitalism uneasily converged on problems of sex's alterability. In other words, this book places feminisms within stories of the wide-ranging political economies and epistemologies which conditioned it.

The women's health movement of the 1970s is critiqued for its narrow rendering of women's health in terms of *reproductive* health—focusing on reproduction not only reified women as simply child bearers; it also so often failed to connect health to racism or larger political economic matters. From a historical angle, this reproduction-focused version of women's health was an important symptom of the period, not just of feminisms in the United States, but also of the moment's larger investments in reordering fertility. It was only in the 1910s that Margaret Sanger coined the term *birth control*, with her politics attaching feminism to Marxism, eugenics, sexology, and professional medicine.[15] *Reproductive health* as a term dates only to the 1970s, later crystallizing as a term of governance in the early 1990s.[16] "Reproduction"—as a political problem and a feature of living-being—itself needs to be historicized within multiple and discrepant genealogies. Reproduction is not so much a "thing" as an overdetermined and distributed process that divergently brings individual lives, kinship, laboratories, race, nations, biotechnologies, time, and affects into confluence. If ever there was a process that is overflowing with contradictory messy genealogies, reproduction is it.

The historian Ludmilla Jordanova points out that, "reproduction" as a term meaning biological generation only dates back to the eighteenth century, arriving into usage alongside the political economy concept of "production."[17] Genealogies of the term show its emergence in eighteenth-century natural history as a way of designating the organization of life within *species*. "Reproduction" in this sense was, rather than a property of the individual, a process of the aggregate, and moreover was a process that re-created an organization of beings out of organized beings. As the historian Londa Schiebinger shows, the modern sense of living-kinds was in turn fashioned in Linnaeus's classification of living things via their "sex," giving us such terms as *mammals*.[18] The work of Stefan Wille-Muller and the economic historian Margaret Schabas reveals that Linnaeus provided one of the earliest descriptions of an "economy" of market exchange at the same time that he offered a description of the organization of life into kinds in an economy of nature.[19] What these historical observations hint at is how *reproduction* as a term through which to organize thought, politics, and life is not at all self-evident; indeed it is the effect of a multitude of genealogies attaching questions of sex and living-kinds to the organization of economics through liberal political thought and knowledge-making practices.

Keeping in mind this knot of genealogies converging on the concept of "reproduction," this book attempts to investigate the feminist politics of reproduction by virtue of historicizing not just the methods of feminisms, but also the very concepts commonplace in late twentieth-century feminisms: woman, bodies, sex, reproduction, and race, as well as freedom, power, and oppression. Extending Simone de Beauvoir's famous assertion that woman is not born but made, feminist projects themselves are assemblages of words, subject positions, objects, and practices each made and not given. "Reproduction" as a target of politics has been repeatedly conjured through varying and uneven distributions of knowledge and practice to produce the historical ontology of sex—woman, man, and child—that has become the groundwork for a tremendous range of projects, including feminisms. So too is the "feminist" as an ethical subject—especially suited to navigating problems of sex—a historically specific summoning at the intersection of political and epistemological concerns.

To claim here that in the late twentieth century a new politics of alterable reproduction was crafted (of which feminisms were a vital part) is not to say that problems of the body and fertility had not existed before. At least since the eighteenth century, since the emergence of liberal feminism phrased within the transatlantic promise of universal citizenship found in the American, Haitian, and French Revolutions, experts and rebels alike have pointed to the anatomies of bodies—of women, the enslaved, or the colonized—as evidence to help decide which humans were human enough to be members of the human universal. In the first half of the twentieth century, heritability, fitness, and racial membership were problems of enormous proportion, trafficking under the name eugenics, cutting a deadly course between nation-states, experts, and ordinary life. What was novel for feminists and other critics in the 1970s, then, was a politicization of bodies that took the means of caring for them, understanding them, and altering them with technoscience as the substance of liberational projects by and for particular groups of people. In other words, feminist projects took up technoscience in order to alter living-being in some ways and not others, investing in an identity politics that named some people (women) as more suited to the ethical management of life. In so doing, feminism was a *biopolitical project*, that is, a project that took life, its kinds and qualities, as the object of its politics.[20]

A century of feminist calls to seize the means of reproduction, to take control of one's own body, to love oneself, to embrace reproductive rights, to end racism, to denounce reproductive technologies, to enjoy sex, to situate bodies intersectionally and so on are all quintessentially biopolitical. Each of these slogans, and others like them, named explicit strategies taken by feminists to concretely do things with the sexed living-being of bodies—including, and especially, capacities to "reproduce." What is there to learn from asking how feminisms took the stuff of living-being—sex, flesh, suffering, pleasure, and especially reproduction—as a prime concern, as phenomena to be rethought and modified?

Historicizing feminisms as a biopolitics that has taken "sex," and its subsidiary "reproduction," as central concerns requires that that we understand feminisms in all their variety and contradiction as animated within—and not escaping from—dominant configurations of governance and technoscience. Since the 1980s, feminist health projects have become one of the most prolific, diverse, and well-funded forms of feminism around the world. Feminist health projects have been able to thrive precisely because they have been so often strategically and uncomfortably conditioned by the financial flows, discursive patterns, and interstices of more dominant configurations of biomedicine, family planning, and economic development. While historians have excavated "the woman question" as a problematic of colonial modernizing projects or have prolifically researched the enmeshment of race and sex within eugenics in the early twentieth century, less understood is the recent past of how "women's health" and particularly procreative capacities constituted an important and well-funded "problem space" of postcolonial formations of nation, empire, race, economy—and of feminism.

Questions motivating this book, then, are: How did reproduction, health, and feminism come to be so intimately connected in the late twentieth century's shadow of American empire? Or by extension, how were local feminist projects based in the United States made possible by larger historical conditions? Or more narrowly, how does feminisms' targeting of reproduction signal the centrality of sex to the emergence of present-day forms of governmentality? I think answering these questions requires unfaithfully rethinking Foucault's initial formulation of "biopolitics," as well as how the history of feminisms in the United States of the 1960s and 1970s, often called "second wave feminism," might be written.[21] How does starting with feminisms rework the traction of "biopolitics" as an analytic?

Foucault's own formulation of biopolitics, which focused on middle-class Europeans, largely foreclosed considerations of colonialism, capitalism, reproduction, or even women. This critique is well known.[22] Given these absences, any account of the history of feminisms as biopolitical in the shadow of American empire would have to reroute the history of biopolitics back through colonialism, the Atlantic slave trade and plantation economies, the calculi of war, the regulation of citizenship, racialized segregation, and formations of global capital, as well as the practices and epistemologies of governance that since Malthus's infamous work on population have connected economic processes and procreation in the many projects of eugenics and population control. Such a rewriting of the history of biopolitics is monumental, yet I think provisionally imaginable as not a single history, but as a discrepant and shifting *biopolitical topology* that helped to yield post–Second World War feminisms and technoscience. In other words, I want to rethink biopolitics: instead of a particular mode of linking life and politics with origins in nineteenth-century Europe, biopolitics is an open question about the manifold ways life became a venue for the exercise of power in a messy, multiterritorialized world.

Biopolitical Topology

In addition to understanding biopolitics as historically situated and plural, I want to reimagine the history of biopolitics as *topological*.[23] Topology names areas of study in mathematics and geography concerned with multidimensional space and crucially with the transformations, deformations, and interconnections within spatialized arrangements. Envisioning biopolitics as topological is useful to thinking historically about the confluence of multiple biopolitical modes at work in any given place within the twentieth century. Beyond just change over time, a topological sense of biopolitics emphasizes: (1) multiplicity, (2) uneven spatiality, and (3) entanglements. In other words, rethinking biopolitics as topological highlights the layered and overlapping configurations that have materialized life in multiple and inconsistent ways over time and across space.[24] Spatializing this multiplicity, then, requires considering how the extension and distribution of biopolitical practices and their effects were profoundly uneven—shaped by race, social movements, nation-states, global capital, segregation, dispossession, urban centers, transnational technical projects, and so on. It is through this topological approach—emphasizing uneven distributions, scales, and multiple layers—that I hope to map the often provincial projects of Californian feminists within larger historical tendencies.

Beyond attending to specificities of scale and time, investigating biopolitics as topological encourages attention to the connections *between* divergently produced instances of biopolitics. In other words, thinking topologically draws attention to the history of attachments, proximities, relationships, fissures, and separations *between* different instantiations of biopolitics. Therefore, methodologically the book strives to go beyond multiplying kinds of biopolitics by focusing on the relationships of appropriation and connection between feminist biopolitics and more dominant forms of biopolitics. It tracks the productive and uneven relationships— antagonistic and supportive, material and discursive—that mutually animated both feminism and other technoscientific practices, particularly in medical and family-planning forms. Hence, the book argues for the importance of attending to *entanglements*, defined as attachments of material, technical, and social relations across divergent and even antagonistic terrains of politics. While genealogy as a method invokes modes of descent, here I attempt to also capture recursive loops, sideway movements, circuits of appropriation, and other vectors of connection within the past.[25]

Such sideways connections can be explicit acts of appropriation between feminism and more dominant technical practices. They can also be points of attachment and exchange that were not politicized or noticed by historical actors themselves. The practices, words, technologies, and subject positions that do the work of attaching discrepant sites are *trans-*

formed as they connect and move in space and time. In the archive, entanglements occur when an abortion device travels between a feminist clinic and a population control program, or when a Pap smear is ethically charged within the walls of a clinical encounter in California and also in a national public health system, and yet again in a transnational safe sex program. Entanglements, then, have ontological stakes as objects and practices are altered as they shuttle between or are shared by different biopolitical tendencies.

Thus, I am not so much interested in cataloguing kinds of feminisms as I am in understanding the emergence of sometimes contradictory feminist health practices through politically laden and layered entanglements. How did feminisms and technoscience discrepantly shape each other through what they appropriated, what they shared, what they disavowed, and what they left unproblematicized?

Despite all the work necessary to complicate questions of feminism as enacted within a biopolitical topology, there is something profoundly useful about the way Foucault initially posed the question of biopolitics as the history of governing living-being, its qualities, kinds, health, rates, deviations, productivities, evolution, and so on. Foucault offered the insight that through a form of often racialized biopolitics, society came to be at war with itself, concerned with the enemy to life within as much as the enemy without. Humans were governed as individual biological beings who were at the same time members of a larger unit: "population," "nation," "species," or "race" — or we might add "economy" and even "women." In other words, as populations were understood to be made up of internal differences, this variation — marked as race, class, pathology, caste, or even sex — could be differentially governed, enhancing some forms of life, neglecting or actively destroying other aspects of life, to bring forth the desired future of that population. Biopolitics thus also always involved necropolitics — distributions of death effects and precariousness — at the same time as it could foster life.[26] It was through this multiscaled *differential governing* of the diversity within the mass, for the greater good of that mass, that individuals in the twentieth century were so often enjoined to participate in the governing of their own potentialities and reproduction. In this way, "population" was not just the ground but the effect of biopolitics, a unit carved in particular ways by demographers, economists, and others that could be used to selectively count and parse life. It is important not to enshrine "population" as a merely numerical unit of living-being that biopolitics is necessarily about, and instead to see the ability to designate *population* as a neutral term—in an era immediately following eugenics—as an *effect* of the exercise of power.[27] "Population" is one aggregate materialized among many others that unevenly enacts biopolitics.

....

Projects that knit together the individual, variation, and an aggregate are not only found in population control efforts of the late twentieth century but also in the many counter-hegemonic identity politics of this period. While "population" in the twentieth century has been sorted to differentially value life through such now canonical categories of race, ethnicity, caste, or class (which in the United States shaped such fundamental features of the period as segregation, citizenship, war, welfare, and market deregulation), related subject positions were also rallying points for political phrasings of many tenors. Identity politics, as a description of a multitude of projects emerging since the 1960s, posits specific aggregate subject positions—such as women—as starting points for politicized counter-conduct. In the case of feminisms, "woman" is both the normative axis of an already-given dominant biopolitical formation, and the founding point for a counter-hegemonic politics that potentially claims all women as its virtual members. For example, through the 1970s in the United States, feminism was widely conceived, particularly by white women, as "by and for women," as a project done by specific, unevenly liberated, female subjects for the sake of Women en masse, a larger collective. Hence the possessive term *women's studies* used for so many academic programs started in this period. The question of what holds together the category "women" in the face of differently situated lives remains a recurrent thematic in feminisms. The contradiction between claiming a universal category "women" while asserting a politics of difference, as the historian Joan Scotts has shown, lays at the crux of liberal feminism and is one of the many constitutive contradictions that has produced feminist politics.[28] This contradictory feature of identity politics, moreover, was just one of the many ways feminisms were fashioned within tangled, contradictory, and tension-filled relationships of a larger biopolitical topography. Hence, it might be useful to historicize the term *identity politics* as an effect in need of critical inquiry.[29] Feminisms within a larger biopolitical topography, took as a starting place the already biopolitically charged subject-position of "woman" within a multiplicity of "women."

What might this reimagined biopolitical topology have looked like as it touched down in Los Angeles in the 1970s, where some of the first self-proclaimed feminist health centers were established? Such a topology would certainly feature the entanglement of life with capitalism—the knitting of capital accumulation with technoscience that occurred in daily life that so many feminists of the period drew attention to.

This observation, however, is too general to be of much use here. While the chapters in the book tell stories about particular technical practices as the grist of biopolitics, here in the introduction I want to lay out some of the broader dimensions of the larger layered biopolitical topology that converged to shape late twentieth-century feminist health politics. As a result, what I offer here is an introductory sketch of a select set of relevant biopolitical tendencies for differentially valuing life via reproduction.[30] These four tendencies are a significant, though not exhaustive, set of animating conditions for late twentieth-century biopolitics in the United States and beyond: the militarization of fertility, the economization of fertility, the industrialization of biomedicine, and the articulation of promissory biocitizenship.

First, after the Second World War, following the Marshall Plan and into the Cold War, the regulation of birthrates in recently decolonized countries became a matter of military concern. In 1959, Eisenhower commissioned a committee, headed by General William Draper, to consider whether United States programs of military assistance—that is, arming select states in the name of protecting the "free world" against "communist encroachment"—were an efficient way of securing capitalist democracy against the "Soviet economic offensive."[31] Draper's committee not only recommended continuing military aid, but equally argued for economic aid to strategic "least developed countries" in order to foster free market economies and the establishment of a single federal agency to distribute this aid (the U.S. Agency of International Development, or USAID). Most controversially, Draper's committee proposed that economic aid alone would be ineffective if the rate of population growth in such countries outstripped production. Poverty bred communism, and birth control was the solution. Only with population control could the United States get "the maximum result out of our expenditure" and achieve military security.[32]

The nuclear bomb as a Cold War weapon of mass death that allowed survival only under threat of annihilation could well be joined by the pill as another icon of the Cold War, used to thwart purported explosive planetary problems of famine, war, and unfreedom caused by the so-called population bomb. With the pill given away as a form of foreign aid, the term *nuclear family* took a militarized turn. The population bomb became another figuration of human mass destruction, seeming to necessitate that the United States both fund family-planning programs along the front lines of the Cold War, and become involved in social science decolonization projects that invented new ways of calculating lives-not-to-be-born as "targets" of population control.[33] In this way, family planning had a particular necropolitical effect—fostering methods for determining lives less worth living in the name of avoiding future death and creating future prosperity. The temporal frame just before and after conception became a new threshold with an important contradiction: it was a moment where human death could be avoided and yet "lives not worth living" calculated and deterred as an ethically charged project. Reproduction was militarized in that family planning could be mobilized to promise a deterrence of future war through its focus on the temporal frame of the "not yet conceived." While the threshold of the not yet conceived was certainly ethically charged, it was a quintessential moment of cold, rather than hot, war, in which militarized violence was displaced and reconstituted under other threats of mass death.

In his analysis of the role of medicine within colonial Algeria during this period, the anticolonialist psychiatrist Franz Fanon argued that even in its very benevolence, medical aid could function as a justification for colonialism; acceptance of health care offered proof that you needed to be saved from your own self rule.[34] While USAID was founded as a federal agency independent from the Department of Defense, foreign aid programs in their many facets could function similarly, as the benevolent face of the Cold War that justified an imperial presence. This further imperial function underlay the tremendous flow of funds, not only into official state family-planning projects but into a new organizational form, the transnational NGO, that helped keep family-planning services work at arm's length from direct rule either by the local state or by the United States.[35] In these ways, investments by feminists in the United States into the management of sex was shaped by entanglements with a militarized imperial history, even when feminist projects were directly antagonistic to population control.[36]

Cold War concern over fertility, moreover, held that the fertility of the world's poor needed to be altered not only in the name of military security, but also as part of a trajectory of economic development. The second animating dominant biopolitical tendency I want to sketch, then, is what I will call the "economization of fertility," the incorporation of fertility into economic planning projects.[37] As the historian Timothy Mitchell and the economist Suzanne Bergeron have both argued, "economy" as an epistemological, social, and technical object only came into prominence as the primary object of state governance in the twentieth century.[38] Even macroeconomics as a field, with its measures of GDP and national inflation rates, only dates to the 1920s.[39] Macroeconomics joined easily with a Malthusian lens, offering ways to calibrate poverty as a natural yet manageable event produced by the conflict between rapid population growth (a biological force) and macroeconomic production. This is precisely how the Cold War field of demography staged the problem of "overpopulation." President Lyndon Johnson offered the pithy synthesis of this ideology to the UN in 1965: "less than five dollars invested in population control is worth $100 invested in economic growth."[40]

The economization of fertility took many forms: overt state population control programs, for example, established first in India and Pakistan, followed rapidly by many postcolonial locales, as well as development projects that declared the status of "women" a crucial point on which economic futures hinged. The centrality of "women" as a pivot of development was signaled by the UN's naming of 1976–85 as the Decade for Women as part of its "Program for Action for a New International Economic Order" of 1974. The fields of demography, population science, and development economics burgeoned, calibrating new quantitative practices and models that often permitted the dollar to be inserted as a unit of measure across economy and fertility. Following the end of the Cold War, Lawrence Summers, then chief economist for the World Bank, influentially argued that women's education was worth investing in precisely because it created good economic returns. He famously calculated that each year of schooling pulls down fertility rates by 5 to 10 percent, such that thirty thousand U.S. dollars spent on educating one thousand women would prevent five hundred births. In contrast, a typical family-planning program that spent sixty-five dollars to "prevent" one birth would accomplish the same for the larger amount of thirty-three thousand dollars. Thus for Summers, "educating girls quite possibly yields a higher rate of return than any other investment available in the developing world."[41] Fertility reduction had become so thoroughly associated with economic productivity that it could now serve as an economic marker for further-removed technosocial correlations.

Not only was the economization of fertility a feature of Cold War and postcolonial governmentalities; it also shaped the biopolitical terrain of the United States. While eugenic targeting of fertility in the name of evolutionary racial futures had shaped federal immigration policies in the first half of the twentieth century, by the 1950s eugenic models of heritability had been scientifically rejected. Demographers critiqued the simplistic biological heredity models of eugenics, morphing social eugenics into social demography that instead held that a "demographic transition curve" charted a population level decline in births as a necessary feature of modernity, and hence that the fertility of populations should now be governed in relation to economic, and not evolutionary, futures.[42] Unlike in thinking on eugenics, racial evolutionary futures were not the focus, though race was still certainly at work in emerging formulations of "cultures of poverty" and designations of who should and should not bear children.[43] In the United States, this ideological change of association between fertility and economics found expression domestically in President Johnson's "War on Poverty." Johnson's program funded nonprofit community centers, staffed by local residents, to offer health, family planning, and other social services, creating a friction-filled privatization of the welfare state, paralleling the proliferation of NGOs in foreign aid projects. Moreover, the Johnson and Nixon administrations' adherence to the fertility-economy equation encouraged an era of state funding for public and especially private nonprofit family-planning programs by organizations such as Planned Parenthood. These programs were further fomented through the reregulation of the management of fertility, resulting in state funding of sterilization through Medicaid, the decriminalization of contraception distribution, and the legalization of abortion. As a result, the tenor of welfare policies directed at mothers reversed direction: single mothers were no longer the deserving poor, but instead economic drains to be removed from the rolls and sent to work as perpetrators of poverty.[44] A popular and racialized logic of economic waste underwrote a period of coercive sterilization in public hospitals, including in California.[45] Economic rationales became the legitimate ground for "choosing" how to manage one's fertility, labeling those who acted otherwise as irresponsible or even failed citizens caught in "cultures of poverty" that therefore needed to be altered.[46] Thus, the economization of fertility in the United States was expressed simultaneously and heterogeneously through the uneven extension of state investment into family planning, racialized economic logics, the retraction of social welfare as a right of citizenship, and the enjoinment of individuals to be economically rational actors open to technical modification.

This investment in family planning was itself joined to a third biopolitical feature of the period: the tremendous changes within medicine itself that observers at the time named the industrialization of medicine, crystallizing by the 1980s into what sociologist Adele Clarke and her collaborators call "biomedicalization."[47] The women's health movement and biomedicalization were contemporaneous, profoundly informing each other. Many of the features of biomedicalization, moreover, were emergent in the 1970s. *Biomedicine* (a term which signals the enmeshment of health care with the life sciences) was exploding as a significant economic venture in the 1970s. The establishment of Medicaid as a national health system for the poor was accompanied by the privatization and corporatization of medicine for the rest. The reproductive and genetic sciences that emerged in this period helped to establish cell lines, embryos, and genetically altered organisms as sources of what Catherine Waldby calls "biovalue," living objects that could be turned into commodities and also used as forms of capital to generate further commodities and services.[48] Sarah Franklin has called this a period of "biological enclosure," where more and more living processes at cellular and molecular scales have become subsumed into capital through their alterability.[49] Charis Thomson, Catherine Waldby, and Sarah Franklin, among others, have demonstrated that such preoccupations with genetics and cell lines—the micrological substrates of sex—rearranged the very terms of capital at the same time that they helped turn "sex" from a problematic, essentialized ground to a flexible zone of artifice.[50] Thus, yet another constitutive contradiction informed feminisms: just as feminists were arguing for a denaturalized conception of sexual difference necessitating the term *gender*, the biology of sex became physically open to reconstruction as itself a changeable domain of life.[51]

This reassembly of capitalism and life in biomedicalization was further accomplished through the growth of the pharmaceutical industry, which offered drugs as widely available commodities manufactured and distributed in new transnational circuits.[52] USAID's underwriting of the global spread of the birth control pill and other contraceptive measures in the 1970s, together with the explosion of family-planning NGOs that distributed and tested drugs and devices, inaugurated some of the infrastructure of today's transnational economy of clinical trials and clinical research organizations and, hence, of the designation of bodies, populations, and even micrological life as sites of biocapital.[53] In other words, reproduction was an important historical locus for the establishment of biomedicalization and biocapital, with feminist health projects formulated in direct and agitated relation to them.

Such agitated relations, moreover, formed the fourth feature of this sketch of manifold biopolitics: the proliferation of nonexpert tactics that sought to render life into governable forms, a process Partha Chatterjee calls "the politics of the governed." For Chatterjee, the politics of the governed are postcolonial projects that self-organize disposessed groups into ethically imbued communities that can serve as the legible target of governmentality.[54] In other words, biopolitical projects are not always efforts to organize others, but can also be projects to self-organize into groups, communities, or identities legible and amenable to modes of governance, including self-governance. In the second half of the twentieth century, the English term *activism* came to denote just such efforts to create counter-conduct modes of organizing life.

Adriana Petryna, in her ethnographic work on how the Soviet state managed the Chernobyl nuclear disaster, coined the term *biological citizenship* to describe how "the very idea of citizenship is now charged with the superadded burden of survival . . . a large and largely impoverished segment of the population has learned to negotiate the terms of its economic and social inclusion using the very constituent matter of life," and in turn states have also been reordered as biopolitical enterprises.[55] The work of Petryna and others has pointed to how so many late twentieth-century biopolitical projects were inflected with the failed promises of citizenship, such that precarious circumstances required the purposeful arrangement of oneself as available for targeting, governance, and technoscientific alteration. Late twentieth-century feminism, in many ways, expresses just such a politics of the governed; it is organized as an ethically charged community that seeks to remake itself by reordering selves. At the same time, this feminist venture of self-making is deeply interwoven with the promises and failures of changing governance during the emergence of neoliberalism. Versions of the women's health movement uneasily inhabited biocitizenship projects that appealed to or worked within the state, while other feminist strategies sought to circumvent the state and reassemble health into sovereign self-governing projects.

Writing in the late 1970s, Foucault's own articulation of the term *biopolitics* was incited by this complex of historical shifts.[56] If liberty seemed to hang on the balance of sex for Foucault's peers, it was precisely because it was made legible, not by developments in the nineteenth century, but by emergent processes in the 1970s better-understood topologically, as having uneven spatial, and not just temporal, extensions in a decolonizing, Cold War world. The politics of reproduction was certainly conditioned by more than these four biopolitical features of the late twentieth century, yet these four tendencies excited each other, forming a shifting topology of connection and rearrangement that gave shape to feminist health practices in California.

Itineraries

This book's title captures its preoccupation with technologies, practices, protocols, and processes—the "means"—of technoscience as crafted by feminist health activists in the 1970s and beyond. Thus, this study is not about how feminists critiqued technoscience. Instead, it focuses on a small set of attempts *to do* feminist technoscience, to fashion feminist biopolitics, in the domain of reproductive health. Since the 1970s, the women's health movement could be found in many sites, enrolling diverse women, expressing various ideologies, founding many projects. This book is not an overarching history of this movement—an important task I will leave to other able scholars. Instead, its chapters center on technologies—the plastic speculum, the Pap smear, and manual suction abortion—as probes that pass in and out of feminisms, tracing itineraries that highlight the differentiating and animating relations between feminisms and other expressions of biopolitics.

My departure point for each of these probes is the radical feminist self help movement of California, a particular fashioning of feminist health care by predominantly, though not exclusively, white lay women of the middle and working classes who themselves were preoccupied with protocols and techniques. Hence, this study is also a lens into the role of whiteness and race in late twentieth-century feminisms of the United States as they were articulated within an emergent imperial political economy and a racialized nation-state, desegregating on the one hand, and inventing new techniques of racial governmentality, on the other. At the same time, the book tells stories about technologies, stories that re-situate California feminist health practices in racial, national, and transnational circulations, stretching beyond the United States. It follows a set of practices and politics as they traveled and became entangled with histories in Barbados, Canada, Shanghai, and Bangladesh, for example.

The first chapter, "Assembling Protocol Feminism," develops the concept of *protocol feminism*—a kind of feminism invested in the politics of technique—and situates the emergence of feminist self help, first, in national and urban racial politics in cities such as Boston and Los Angeles and, second, in the rise and dissemination of Cold War small group techniques of human relations research.

Chapter 2, "Immodest Witnessing, Affective Economies, and Objectivity," tracks the epistemological experiments around clinical exams through the plastic speculum, and places these practices in the larger history of scientific objectivity, as well as the elevation of "affect" as a virtue within feminism and feminized labor. In so doing, it develops the notion of *affective economies* of knowledge.

"Pap Smears, Cervical Cancer, and Scales," chapter 3, maps divergent politicizations of the ubiquitous Pap smear over the second half of the twentieth century, tracking how discrepant feminisms have scaled the problem of cervical cancer in clinics, national screening programs, and transnational health policy. This chapter attends to the relations of appropriation and reappropriation that entangle variously scaled feminisms with biomedicine, racial governmentality, and transnational economic development logics.

"Traveling Technology and a Device for Not Performing Abortions" chapter 4, plots the entanglements between feminist attempts to do abortion differently and transnational population control, highlighting the various ways "freedom" was hinged to reproduction through both feminism and family planning sponsored by the United States government.

Finally, the conclusion, "Living the Contradiction," builds on the insights of these four chapters to think through the importance of attending to the work of contradiction in these histories. Overall, the book puts into play "woman" as the assumed or sufficient subject of feminism, and feminism as a frame for reimagining new technoscience futures.

W. E. B. Du Bois famously described a "double consciousness" that arose from the contradictions of being a member of the "problem" that he was studying.[57] This book attempts to work another instance of double vision: ruthlessly historicizing these past feminist efforts as one might any other scientific endeavor, while doing so from a point of deep investment in feminist technoscience studies as a critical epistemological and material project that values entanglement and sits in a genealogic relation to the practices examined. Gayatri Spivak elegantly noted that deconstruction as an intellectual project was not driven by a concern with exposing other people's errors but instead sought to constantly and persistently look at those things without which one cannot live.[58] It is in this spirit that I seek to historicize feminism, technoscience, and reproductive health.

In the years since these feminist experiments with doing health differently, many of the terms initially mobilized here have gone on to follow complex and discomforting itineraries, out of marginal radical projects into World Bank or World Health Organization guidelines, state policies, and national research agendas. *Participation* has become a buzzword for structuring development projects in such a way as to require the involvement of the people whom they target. *Empowerment* has become a technocratic goal that directs the flow of resources and training down the chain from prosperous to more precarious NGOs. *Gender* as a term has not only been repeatedly redefined in circuits of linguistic and disciplinary translation, but has become an organizing spoke of the World Health Organizational and U.S. National Institutes of Health. Forty years after 1970, reproduction's alterability is no longer a promise, but instead has become a normative condition, such that the *inability* to manage reproduction is reframed as *a product* of the uneven extension of medical services and rights across the globe.

While the particular moment of United States feminisms in the shadow of the Cold War and postcolonial politics is now past, it sits as an important prehistory to the ways health is governed and politicized today. Thinking feminism as biopolitics lies at the heart of this book's iteration of feminism, as does the question of whether the contours of feminism are sufficient to that project. What kinds of ontologies can feminisms and technoscience excite or foreclose when "woman" is assumed as a privileged ethical subject? It would be a mistake of this book (that I have struggled against) to present the question of feminism as biopolitics in terms of failure, or more simply to equate biopolitics with exploitation, forsaking Foucault's injunction to understand the exercise of power as productive. Thinking feminism as biopolitics is also about yearning to continue experimenting with technoscientific practices that could foster better means of enabling life with eyes open to the constitutive contradictions of an entangled world.

CHAPTER 1

Assembling Protocol Feminism

A feminist self help clinic in 1970s California might be found at a local Women's Center with participants perched on shabby sofas below a poster of a raised fist clenching a speculum. Or it might be held in a home with children and spouse tucked away, or even in a church basement. Self help clinics could form in any nonmedicalized setting, with women examining themselves and each other on couches, chairs, or pillow-topped tables, as much as at a formal feminist health center. No sterile blue paper gowns or obstructing medical drapes were required. Instead, participants wore street clothes, taking off skirts, pants, and underwear, but casually leaving on socks and knee-highs (see figure 1.1). While a self help kit's iconic tool was the cheap plastic vaginal speculum, it might also have included information on local abortion laws, a mimeographed list of local abortion clinics, the twenty-five-cent *Birth Control Handbook* from Montreal, or instructions for starting your own advanced research project.[1]

"A Self Help Clinic is not a place," so feminist self help founders asserted. "It is any group of women getting together to share experiences and learn about their own bodies through direct observations."[2] In other words, a self help clinic was not locatable within the physical walls of a medical facility. Instead, it was a mobile set of practices, a mode for arranging knowledge production and health care, in other words, a *proto-col—a procedural script that strategically assembles technologies, exchange, epistemologies, subjects, and so on.*[3] Put simply, a protocol establishes "how to" do something, how to compose the technologies, subjects, exchanges, affects, processes, and so on that make up a moment of health care prac-

1.1. A photograph used in the slide show by the Los Angeles Feminist Women's Health Center in the early 1970s. Here, casual dress, living room furniture, friendly demeanors, and mundane commodities are assembled into a feminist self help clinic. Courtesy of Lorraine Rothman.

tice. A feminist self help clinic provided choreography for "how to" assemble sexed living-being, technoscience, and politics.

Self help clinic was the official term used for events—not places—organized by Los Angeles area Feminist Women's Health Centers in which women, mostly but not exclusively white, and often strangers to each other, met for a set number of weeks to learn self-examination techniques associated with gynecology and to "demystify" their own bodies as facilitated by a lay health worker. Visual examination of genitals and cervixes was joined by tactile techniques of palpating uterine size and position. Protocols of examination, moreover, were performed along with social protocols of "consciousness raising"—women sitting in a circle on the floor comparing experiences and observations, "as we did then."[4] Conventional medical methods were designed for anonymous encounters between doctors and strangers, while feminist self help was to be practiced by "a cluster of women" who had earned an intimate and affective knowledge of each other's bodies.[5] It was crucial to feminist self help that it not be a solitary practice—it required a group to instantiate acts of "care of

HOW TO DO SELF EXAMINATION USING THE PLASTIC SPECULUM
ⓒ Copyrighted by Self-Help Clinic One 1972

Lubricate your speculum with the duckbill closed and the handle in an upward position. Grasping it by the duckbills, gently insert it with the bills closed into your vagina, as you would insert a tampon until the handle touches the pubic area. (Many women prefer to insert the speculum sideways and then turn the handle up.)

Do not insert the speculum if you experience pain. You may need a smaller size. Please let us know and we will send you a smaller-sized one.

To see yourself, hold the mirror between your legs and direct the light toward it. The light will reflect off the mirror into your vagina so that you can view your cervix in the mirror. Do not be discouraged if you are not successful your first try. The speculum may have to be moved around or be reinserted before the cervix will pop into view. Having someone with you to tell you when it is in view is of great help. Many women have had to try several times before they were successful. But they were _always_ successful.

Squeeze handle together (this will open the speculum within your vagina). Press down the part of the handle with the finger depression while pulling up on the longer handle. This will lock the speculum open. You can adjust the handle to three positions (you can hear it click in place).

Feminist Women's Health Center
Self-Help Clinic
429 So. Sycamore
Santa Ana, Calif.
92701

714-836-1941

After viewing yourself, remove the speculum, still open, by slowly pulling it straight out. Wash it with hot water and antiseptic soap and store in a clean place.

Feminist Women's Health Center
Self-Help Clinic One
746 Crenshaw Blvd.
Los Angeles, Ca.
90005

213-936-7219

1.2. An early and typical handout explaining the protocol of vaginal self-examination. From the Boston Women's Health Collective archives.

the self" into politicized and experimental modes of nonprofit exchange ("sharing") and knowledge production ("consciousness raising").

These protocols of self help traveled broadly across the United States and Canada through a flurry of mimeographed, and later photocopied, how-to flyers and pamphlets that described, in text and pictures, instructions for forming a "self help group" (see figure 1.2). Sometimes, after an organized clinic ended, women would continue meeting in their own "advanced groups," moving from basic examination to the articulation of more avowedly experimental projects, such as investigations of sexuality and female ejaculation, lesbian health, or the practice of "menstrual extraction."

Donning white lab coats to surreptitiously purchase urine pregnancy test kits (then medical supplies administered by doctors, not over-the-counter commodities) the inaugural self help clinic began by teaching women how to conduct urine pregnancy tests on themselves at the local Women's Center in a house on Crenshaw Boulevard in Los Angeles.[6] Even later, within the formal women's health centers that provided abortion and reproductive health services—which rapidly emerged after 1973 in Los Angeles, Orange County, Santa Ana, Santa Cruz, San Diego, Oakland, Chico, and elsewhere—the minutiae of practice was profoundly politicized.[7] How health histories were taken (as "herstories"), insistence on group alternatives to individual doctor–patient encounters ("participatory clinics"), the requirement of narrating abortions as they were performed, and even the exact instruments chosen were all open to politicization and revision.[8] At its peak from the early 1970s to the early 1980s, between the apex of radical feminism and the rise of militant Reagan-era antiabortionism, these scattered projects formed a reticulate, experimental, and influential hands-on strand of a nationally interconnected women's health movement.[9]

Feminist protocols were typically designed to spread and be mobile. They were often transmitted by live demonstrations first initiated by a handful of Californian women who undertook "road trips," traveling by station wagon or bus and sleeping on couches to arrive at college towns, big cities, and church basements, where they gave slide show presentations culminating in a live performance of vaginal self-examination. This labor was privately referred to as "PR" work, accomplished not only by travel and text, but also supported by the assiduous documentation of practices through home movies, photos, and slides that increasingly portrayed a multiracial constituency of participants. Thus, the expression and spread of feminist self help protocols was dependent on cheaply available technologies of popular photography and photocopy reproduction, as well as on infrastructures of highways and bus systems. Such infrastructural possibilities were additional ingredients in the feminist reassembly of the terms of health care. Sketches, photos, and instructions from paper flyers made in one town were cut, pasted, and recopied in another to create local versions of feminist self help. By the mid-1970s, feminist self help had been demonstrated at the UN Conference on Women in Mexico City, while projects linked by their citations to each other were established in Germany, France, Britain, Canada, Australia, New Zealand, Barbados, India and Brazil, to name just a few sites in a transnational itinerary.

In foregrounding the politicization of techniques and practices as a means of also designating "women's health" "bodies," and "sex" as politically charged sites, feminist self help, I argue, was a kind of *protocol feminism*—a form of feminism concerned with the recrafting and distribution of technosocial practices by which the care and study of sexed living-being

could be conducted. In other words, the politicization of life in the formulation of "women's health" was simultaneously bound to the politicization of the details of the techniques by which it was known and materially altered. While feminisms are often categorized by historians into
the slots of liberal, postcolonial, or Marxist (for example), such templates
fail to capture the range of feminist counter-conduct—such as feminist
biomedical and policy projects—that have multiplied in the last forty
years, as well as the myriad feminist formations—such as entrepreneurial
feminisms or imperial feminisms—that shore up and lend ethical legiti-
macy to dominant formations.[10] A daring form of counter-conduct in the
1970s, protocol feminism has since become a common mode of feminism
that can today be found in NGO-ized projects and bureaucracies concerning health, international development, and domestic violence, as well as
international policy.

Protocol is a term used widely in biomedical practice to name the
formal guidelines, instructions, or standards for composing a task—all
the steps in drawing blood, for example, from assigning staff, to standardized orders, to attaining consent, to syringe angle, to labeling, to dis-
posal, as well as all the choreographed arrangement of subject positions
and institutional players it draws in—doctors, nurses, hospital management, lab technicians, insurance agents, medical device companies, and
of course patients.[11] Feminist self help, as a protocol feminism, likewise
assembled together bodies, feelings, tools, modes of politicization, social
interactions, relations of exchange, and emerging biomedical logics converging on questions of reproductive health in the 1970s. Unlike medical
protocols, offered as rational and apolitical technical achievements, feminists saturated protocols with politics.

Feminist self help did not emphasize the term *protocol*, but instead
talked of process, structure, procedures, and practice. Turning to the term
protocol here helps to highlight the standardizable and transmissible components of feminist practices. Moreover, the question of protocol draws
attention to the scripting of *relations* between component entities—the
instruments, labor, gestures, identities, emotions, and so on—assembled
to compose feminist practices. Here, I am building on the insights of theorists Gilles Delueze and Felix Guattari, who describe an assemblage as
composed, not by the list of tools and components, but by the intermin-
glings that make the tools possible.[12] In other words, it is not the historical
availability of a set of components which assembled feminist self help; it is

the arrangement, composition, or *protocols* which actualized the elements in some ways, not others, evoking historically specific generative capacities to act, to matter, to care, to count or be counted, to attach, to emote, to narrate, to ignore, to work, to value, to politicize, and so on. *Protocol* is a word that describes the specific choreography of such evocative relations.

At the same time, feminist protocols were always a *reassemblage*, not simply historically new but crafted by appropriating and altering elements already available in the 1970s. As a reassemblage, feminist self help was entangled with, and did not wholly repudiate, the historical conditions of its emergence. This state of reassemblage is more than just a historiographical observation of this book; it was also an explicit tactic embraced within feminist self help. On a concrete level, feminist self help's two iconic artifacts—the plastic speculum (featured in the vaginal self-exam, detailed in chapter 2) and the menstrual extraction kit (of chapter 4)— were examples of reassemblage, one reconstituted from a device used in conventional medicine, the other made by rejigging an instrument from an illegal abortion clinic with a mason jar and aquarium tubing. Feminist self help was thus shaped by an announced ethic of reassemblage in the sense that its practitioners were often tactically attempting to appropriate and reshape existing practices or technologies.

Even further, feminist self help emphasized its own internal ethic of *flexible and experimental* reassembly, aspiring to craft protocols that could foster change, move between sites, and be tailored to particular needs as decided by individuals or small groups. Feminist self help focused on practices that could be done by lay people routinely, suitable for a living room, and constructed out of common items. Since feminist self help practices were rooted in a reflexive and collective project of investigating one's own body, its protocols ideally were always under revision to suit local politics and the particularities of individual embodiments. Through this explicitly experimental ethos, the protocols of feminist self help were multiply revised as they moved across time and place. And then later in the century, this quality helped to facilitate the reappropriation of feminist protocols within conventional biomedical, family-planning, and development practices. Thus, feminist protocols served as more than instructions for assembly; they actively encouraged revisable and mobile appropriations and rearrangements.

Though marginal in the 1970s, today protocol feminism has become a common element within contemporary biomedical and developmental modes for governing sexed life, in which feminist and other politicized projects—typically within NGOs but also within states—aspire to intervene in and ethicize the standards, terms, policies, and guidelines through which women's lives are governed. Examples range from feminist NGOs that write manuals of best practices for family planning, to concerns over the "process" of meetings, to feminist efforts in the early 1990s to change the language of UN population policy to that of "reproductive health."[13] Moreover, the project of feminist technoscience studies (of which this book is a part) is in kinship with protocol feminism as expressed through its promise of doing science another, better way.

In attending to the importance of protocol within Californian feminist self help of the 1970s, I want to insist on two claims from which to historicize the traffic between feminisms and technoscience. First, I want to argue the broad claim of this book that the histories of feminism and technoscience are conjoined, and that this relationship is not merely a story of feminism's critique and correction of science. Instead, practices such as feminist self help were examples of technoscience in the twentieth century, even as they were in agonistic relationships with other forms of technoscience. The proliferation of diverse forms of participatory science and politicized health and environmental projects in the late twentieth century is as important a feature of technoscience as were particle accelerators, industrial labs, and the rise of biotechnology. Feminist self help was a politicization of technoscience as much as of reproduction.

Second, I argue that this study of feminist self help requires the historicization of feminisms through the same methods one would use to treat any other instances of technoscience. Rather than an account of successive, ever improving, waves of feminism, or of separate roads to feminism plotted by different racialized communities of women, or a story of the errors of feminism and its failures to accurately prophesize neoliberal futures, this chapter seeks to chart the assembly of feminist self help in Los Angeles in the early 1970s as a protocol feminism made possible by discrepant histories of politicizing life and health.

Feminist self help as a protocol feminism drew together and reacted to already extant practices, infrastructures, commodities, epistemologies, and subject positions available in the landscape of its emergence. I will leave detailed questions of the epistemological content of feminist self help to the next chapters, and the social history of the women's health movement to other scholars.[14] Here, I am interested in tracing some of the entanglements—the acts of reassemblage, appropriation, and disavowal—that variously agitated and animated Los Angeles feminist self help as a protocol feminism within larger historical conditions of possibility. By following a nonexhaustive and select set of genealogical entanglements as they relate to the politics of protocol, this chapter treats protocol feminism as complexly conditioned by discrepant relations outside itself and at the same time as a reassembly of those conditions

In other words, feminist self help was both a *symptom and diagnosis* of its moment. Feminist self help critically *diagnosed* and redirected the exercise of power as it moved through technical practices that invested reproductive health with new political dispositions. At the same time, it was a *symptom* animated by and entangled within both existent and often contradictory conventions of stratifying and politicizing living-being (such as through nationalism, race, citizenship, labor, and so on) as well as emerging milieus of technoscientific practice affiliated with the Cold War, new racial logics, and capitalism.

In striving for this double vision of feminist self help—as both made out of and making a larger biopolitical topology—I do not want to lose track of why I first became interested in the history of feminist self help in the first place: it was a practical and influential instance of an attempt to fashion a feminist technoscience that has materially shaped my life and those of many others around me. It was formative to the epistemological and political investments of my own scholarly field of feminist technoscience studies, itself crafted in the 1980s.[15] Feminist technoscience studies, in turn, has been a critical aspect of the field of science and technology studies more generally, as well as more recently of the field of women and gender studies. Thus, I want to invite interested readers to consider how science and technology studies or women's studies, which emerged simultaneously to feminist self help, are likewise imbricated in this historical formation.

The bulk of this chapter is concerned with charting four histories that inform how feminist self help as a protocol feminism was assembled: (1) the new legibility in the 1970s of "biopolitics" as a form of politics connected to technoscience, (2) the often racialized designations of healthful and precarious lives that shaped the ways various feminists theorized the politics of subjectivity in their projects, (3) the dawn of biomedicalization and establishment of population control and large-scale family planning, which was accompanied by a corresponding shift in the politicization of health, and (4) the politicization of small "group" practices as part of both decolonization and democracy in the Cold War and postcolonial logics of the period. In doing so, this chapter has yet one more goal: to introduce some of the overarching elements of feminist health practice that will recur in the rest of the book.

Historicizing Biopolitical How-To

West Coast feminist self help declared their tactics "revolutionary," providing the means to "take control of our bodies" or "seize the means of reproduction," or "take back turf."[16] Criticizing medicine as a patriarchal and profiteering profession, feminist self help sought to design *practices* by which reproductive health care could be conducted outside of biomedicine proper, outside professionalism, and outside of legal regulation, as well as outside of profit imperatives.[17] From the act of vaginal self-exam, to the details of abortion technique, to the administration of a clinic, to the exchange of information, to interpersonal conduct, to the vocabulary for illustrating bodies, self help feminists were explicit about their focus on the micrological politics of health care *practice*. Primarily practices, not questions of access or diagnosis, were their foci, and it was practices that they theorized as profoundly imbued with politics. This focus on "practice" aligned feminist self help with rape crisis centers, battered women shelters, and other radical feminist projects of service provision that tried to prefigure the social relations wished for in the microcosm of an organization. Unlike service provision however, feminist self help was theorized as a moment of self-sovereignty, not as labor for others. According to self help doctrine, no expensive instruments, white coats, or prestigious degrees were necessary for basic gynecological health care; all you had to do was use your body to study and care for your body. Actions of self-study were theorized as a means to take "control of our own bodies," as expressed in this familiar credo of the movement.

Within radical feminism, the call for women to take control of their bodies was inherited from Margaret Sanger and the birth control movement of the early decades of the century.[18] Sanger, then a young member of the Industrial Workers of the World (Wobblies) and involved in the Lawrence Mills Strike of women textile workers in Massachusetts, in turn cribbed the phrase from Marxist socialist visions of a revolution in which the proletariat "seized the means of production." Later, Sanger would bind this call to eugenic projects. However, the Marxist inheritance carried well into the 1970s, such that Claudia Dreifus, editor of the book *Seizing Our Bodies* (1977), declared, "It is not factories or post offices that are being seized, but the limbs and organs of the human beings who own them."[19] The Santa Cruz Women's Health Collective joined the call to take back control with the slogan "Health Care is for People, Not Profits."[20]

In addition to building on socialist ideology, feminist self helpers also appropriated anticolonial rhetoric. "Women are a colonized people, with our history, values, and cross-cultural culture having been taken from us—a gynocidal attempt manifest most blatantly in the patriarch's seizure of our most basic and precious 'land': our bodies," wrote Robin Morgan, the New York radical feminist and poet in the introduction to a Colorado guide (1975) to feminist self help. "Our bodies have literally been taken from us, mined for their natural resources (sex and children), and deliberately mystified. . . . We must begin, as women, to reclaim our land, and the most concrete place to begin is with our own flesh. . . . Identification with the colonizer's standards melts before the revelations dawning on a woman who clasps a speculum in one hand and a mirror in the other."[21] While reassembling anticolonial and Marxist discourses to critique the appropriation of women's reproductive capacities by "patriarchy," feminist self help in the United States typically remained insensible to the paradox of a rhetoric of domination at work in their own calls to take individual possession of their flesh.[22]

In its wide use, the call for women to take control of their own bodies was extremely adaptable to a variety of feminist stances; Chicana and black feminists in the United States used the phrase, which was then redeployed in transnational feminist assertions of a liberal "reproductive right" to bodily integrity and choice.[23] For example, many calls for "reproductive control" on the part of American black feminists situated it as a long historical struggle continuous with resistance to enslavement, thereby seeing bodily control in relation to civil rights or political economies of racism and incarceration.[24] West Coast feminist self help, in contrast, tended to theorize control as appropriable through the micropolitical details of techniques for directly acting on one's own the body. Their micropolitics of technique was oriented toward fostering an individual and embodied sovereignty of "choice" that hoped to function as a flexible, "nonjudgmental" ethic that avoided setting "prescriptions." In so doing, *reproductive control* applied to the freedom to bear and not bear children, thereby aspiring to include within its protocols an individualized opportunity to circumvent coercive and racist measures, yet without directly theorizing racism. It was the achievement of self-sovereignty, not what you did with that control, which mattered to this protocol feminism.

Within the feminist self help movement, reclaiming one's body from patriarchy was not meant to free a natural body from the grip of culture or artifice; feminist self helpers did not romanticize the experience of unwanted pregnancies when birth control and abortion were illegal. In this way, they ideologically differed from, for example, those feminist midwives who called for "natural" childbirth. Instead, taking back "control" was a matter of asserting an *active* relationship to one's own biology. It was intended as an assertion of sovereignty over oneself—of *self-possession*—enacted through practice. Hence, it was the "means of reproduction," the practices and technologies used in the management of reproduction, that were seized. And it was protocols that recaptured these practices and techniques, drawing them into new orders. Seizing the means of reproduction, moreover, required an ethic of *operability*—a term the anthropologist Lawrence Cohen uses to describe how people make themselves available to biomedical manipulation for the sake of participation and recognition within nationalism, modernity, reason, or, in this case, feminism.[25] Thus, "seizing the means" involved politicizing the techniques of this operability as self-possession.

Historicizing Californian feminist self help as an experimental protocol for doing technoscience differently provokes questions about not only the history of its technological assembly, but also about the history of politicization—what counted as "politics" and where politics was seen to be in operation. Here, again, I want to offer a double view, where I attend to both politics analytically as the exercise of power immanent in all practices, and hence all technoscience, as well as politics as a historicizable announcement that identifies some practices and domains as political—in other words, *the politics of "politics."*[26] Feminist self help, for instance, imbued the minutiae of medical practice as politically charged, and hence sought to appropriate this power immanent in the details of technical practice. In so doing, feminist self help simultaneously and explicitly imbued "women's health"—a demarcated domain of life on which techniques acted—as a site of politicization. In this sense, feminist self help announced the legibility of life and technoscience as two politicized and entangled domains—as what we can analytically call biopolitics and technopolitics.

Foucault famously developed the term *biopolitics* in the 1970s to describe the historical emergence of a form of governmentality (understood as techniques for directing conduct in sites that can exceed the state) which took living "populations," particularly sex as its targets.[27] Moreover, biopolitical practices did not just take life as a political concern; they also gave form to new ways of apprehending, ordering, and manipulating life. His early work on biopolitics suggestively hinted that the politicization of sex as a form of liberation in movements of the 1970s—both gay and feminist—was only the latest chapter in a longer biopolitical genealogy.[28] Declarations of liberation from repression, according to Foucault, were in fact symptoms of the production of sex, rather than escapes. Moreover, Foucault's own theorization of power as exercised—that is, inherent in the specific relations generated by practices, instead of a quality possessed by individuals or institutions—drew attention to the microdetails of practice as the constitutive register of politics.[29]

Just as today the recent past of late twentieth-century feminisms has moved into the ambit of historical analysis, so too has Foucault's analysis, which dates from the same period, become increasingly historicizable. That he was asking about biopolitics and micropolitics in the 1970s, and that the term *biopolitics* has become an important topic of study today, suggests that biopolitics might be particularly evocative of late twentieth-century and contemporary rearticulations of life, rather than primarily a description of developments in the nineteenth century as outlined by Foucault. Thus, it is possible to think doubly about biopolitics—first as an analytical term that helps us to excavate histories of practices through which living-being was governed and, second, as a term in need of historicization that was articulated in the 1970s at a moment when sex, race, capitalism, and health were repoliticized within radical projects.

Feminist self help as a protocol feminism, then, can be understood analytically as a radical and marginal mode of biopolitics, in which individualized living-being (bodies) and an aggregate living-being (in this case, "women") were reknit in the 1970s as an explicit political project. At the same time, feminist self help can be historicized to ask how it became legible to feminists, as it had to Foucault, that the minutiae of techniques was constitutive of politics. This approach to thinking doubly about biopolitics does not involve faithfully applying Foucault's analysis of the nineteenth century to the 1970s; instead, it involves rephrasing questions of biopolitics as newly enunciable in the 1970s. As an example of the "politics of politics," biopolitics names life as the legible domain of politicization, yet still leaves open the question of how politics was itself mapped and given tactical shape. For the feminist self helper, as for Foucault, the legibility of biopolitics in the 1970s was articulated through a sense of technopolitics—a sense of the immanence of politics in practices and technologies.[30]

This question of the politics of politics is at stake in the rest of the chapter, as it turns to the task of more concretely outlining three of the divergent milieus from which feminist self help was assembled as a protocol feminism. The questions of how to politicize subjectivity, health, and group sociality were vital to feminist self help formulation of its protocols. Moreover, the work of "race" was integral to all these versions of the politics of politics.

Race, Necropolitics, and the Politics of Subjects

Feminist self help's moment of radical counter-conduct flourished between 1970 and 1980. It is in this decade of deindustrialization, white flight, and proliferating counter-conduct preceding the inauguration of state neoliberalism under Reagan, that feminist self help as a biopolitical project was deeply informed by *whiteness*. By whiteness, I do not mean a biological property of bodies, but a racialized social formation of citizenship, embodiment, and economy that materially arranged the lives, homes, labor, vulnerabilities, and epistemologies of the subjects it both hailed and refused to hail.[31] In other words, whiteness is not a property of persons, but a historical formation that has differentially charged subjects as available for privileges, property, injury, and dispossessions through their bodies. If biopolitics is analytically understood as strategies for politicizing the bind between individual bodies and aggregate forms of life, the "we" that Los Angeles feminist self help announced calls for historicization on a wider terrain of whiteness and the feminist-conjured aggregates—the many versions of "women" raced and unraced—provoked by political economies of late twentieth-century America.

In the United States, whiteness largely operated as unraced and unmarked to its beneficiaries, shaped by the historic norm of white male citizenship. Whiteness could stand in for a liberal ideal of a self-governing individual who ideally transcended his or her conditions, such that social disadvantages did not mark them. Feminist self help's version of self-care and individual bodily control also presupposed a self-determining, self-knowing, self-possessing subject-figure as attainable and universalizable. Yet, this version of protocol feminism reassembled the liberal sovereign subject (presupposed masculine) of "possessive individualism" by virtue of sexing and collectivizing it through shared practices with other women.[32] The small self help group heralded this self-possessed subject through technical acts on the body, binding the figure of the self-determined person, not to nation or capitalism but to a "community of women" or "sisterhood" as its condition of possibility. Here the biopolitical *we* was demarcated through visions of unraced, unclassed sex.

Unspoken by most feminist self help practitioners in California and elsewhere, then, was the work of race and class formations in these biopolitical maneuvers. The efflorescence of feminist politicizations of health in Boston provides a useful foil through which to juxtapose the work of whiteness in the sexed subject-figure of Los Angeles feminist self help. Boston in the 1970s was an older city, watching factories close and a biomedical economy grow. It was an intellectual hub with the nation's highest concentration of public and private universities, and at the same time a city split by class and racial segregation that showed itself, for example, in the racist resistance to school integration through bussing during this period.[33] Hence, the politicization of race took a different form in Boston than in Los Angeles, where in contrast racial and class stratigraphies were shaped by more recent settler colonization of Mexican and Native American territories, as well as Latin American and Asian diasporas.

Boston was at once home to a stunning variety of feminist health politics: the Boston Women's Health Book Collective that published *Our Bodies, Our Selves*, which initially suggested that "learning about our womanhood from the inside out has allowed us to cross over the socially created barriers of race, color, income and class, and to feel a sense of identity with all women in the experience of being female";[34] the Somerville Women's Health Clinic, which was founded by a collaboration of local working-class women and middle-class health professionals and defined itself as serving the "community," working-class women, men, and children; and the Combahee River Collective, authors of the influential (and now canonical) "A Black Feminist Statement."[35] Today taught regularly in women and gender studies curricula across North America, the statement crafts a complex analysis of "manifold and simultaneous oppressions" that necessitates multiple coalitions to address the "multilayered texture" of "interlocking" systems.[36] While the "Combahee River Collective's Statement" is credited with coining the term *identity politics*, it is crucial to note that it does not affirm a single epistemically privileged identity or name an authentically revolutionary subject position, but rather draws out the contradictions formed at the axis of race, class, sexuality, and gender, as well as logics of capitalism and state violence, that require *disidentifications* from any singular identity and thus the recognition of contradictory difference.[37]

That the politics expressed in the "Combahee River Collective's Statement" was further forged on a landscape of health politicization is revealed when the statement lists the "issues and projects that collective members have actually worked on" as sterilization abuse, abortion rights, battered women, rape, and health care.[38] One founding member of the collective was Beverly Smith, who taught one of the earliest courses on "women's health" at the University of Massachusetts, Boston, and developed a syllabus explicitly focused on black women's health.[39] Smith's concern with questions of feminist protocol was signaled through her master's thesis in public health: "The Development of an Ongoing Data System at a Women's Health Center."[40]

The center in question was the Women's Community Health Center of Cambridge (WCHC), formed in 1974, partially inspired in 1973 by a feminist self help presentation held in Worcester. In its own reassembly of the feminist self help manifesto, the WCHC proclaimed, "We see what we are doing as a political act; it is therefore essential to define ourselves politically. The following is our collective statement . . . We feel that sexism, patriarchy, capitalism, and racism are inextricably intertwingled."[41] Tellingly, WCHC—with its feminist, antiracist protocol—was attracted to the word *intertwingled*, a neologism of Ted Nelson's, an early Boston internet "pioneer," author of the 1974 photocopied booklet *Computer Lib*, and coiner of the term *hypertext*, which like *interwingularity* was an attempt to capture the intermingled aspect and "cross-connections" of knowledge.[42] Foregrounding "intertwingled" systems of oppression, the WCHC statement described feminist self help as a tactic that could be agreed upon across its members' divergent ways of articulating feminism. Thus, WCHC crafted its "politics" in both direct relation to questions of information protocols—databasing and hypertext—and to analyses of interlocking oppressions by Boston-area women of color. The phrasing of political intersectionality through a term describing software—*intertwingularity*—was particular to the Cambridge articulation of protocol feminism and its investment in technoscience with tendrils extending to both MIT and the Combahee River Collective. What this juxtaposition of Boston-area feminist projects suggests is not only the variety of feminisms between and within cities, but also the varied maps feminists created of the biopolitical stakes of their project, that is, the varied ways feminists politicized *bios* and *techne*.

The Los Angeles version of feminist self help, with its focus on the "healthy" woman and the achievement of self-possession, thus, can be situated as both a diagnosis and a symptom of a larger uneven topology of sexed and raced biopolitics within America. Here, I want to think not of a history of separate feminisms simply sorted by race; rather, I want to conceive of a refracted traffic of feminisms entangled and yet discrepantly realized. Feminisms were shaped by specific local urban instantiations and distinctive distributions of both injunctions to health and vulnerabilities to harm, all profoundly racialized. Los Angeles feminist self help can be understood, not as the temporal point of origin to which all other versions of feminist self help are indebted, but as one announcement of feminist biopolitics among a larger proliferation of near simultaneous feminist articulations. As such feminist self help, despite its Federation of Feminist Health Centers along the West Coast, was a decentralized and variegated formation, the exact expression of which was ideally, and in practice, reassembled and repoliticized in each local instantiation.

Historicizing protocol feminism within a larger topology of discrepantly expressed racial politics opens questions about the ways whiteness and the selective recognition of race operated in the emergence of feminist self help in the particularities of Los Angeles. Founders of the first LA Self Help Clinic described themselves as "six white housewives who had 24 children among us."[43] It might be tempting to call Los Angeles feminist self help a "white feminism" that can then be compared to "Black feminism," "Chicana feminism," or "Asian feminism," all of which found expression in LA. However, the mostly white women who held leadership positions in Californian feminist self help clinics of the 1970s did not mark their projects under the banner "white." Unlike those feminists in the early twentieth century who explicitly named themselves "white" as a means of marking feminism within an imperial civilizing project, in the 1970s *white* had a pejorative ring, at least on the Left, and was rarely uttered as a positively politicized self-description.[44] In a decade when feminisms in the United States were so often bound with racial or ethnic formations and allegiances, the types of feminist projects practiced by predominantly white women that did not explicitly hold themselves accountable to addressing race might be usefully called *unraced feminisms*. Such projects were not only silent on questions of race; they tended to shore up the unmarked and normalized work of whiteness by unracing themselves.[45]

It has become a truism for critics and historians of late twentieth-century feminism in the United States that many feminist projects organized by predominantly white middle-class women identified their politics under the generalized unraced category of "women." The historical scholarship on late twentieth-century feminisms, however, could do more than identify this "error" and ask of these projects: How did different formations of racialization assemble in discrepant feminisms? I want to think this question through more specifically by continuing the juxtaposition between Boston and LA feminist politicizations of health.

Feminist self help often strategized itself as stripping away false pathologization and "mystification" created by medicine to reveal the authentically normal and healthy woman within. Whiteness, then, worked in unraced feminist self help to constitute a particular vision of the subject—the flexible and abstract, "every woman" who was hailed as "healthy," "normal" and individually "unique," who, moreover, could achieve self-possession and normality. In this way, every woman functioned much like the figure of the abstract citizen: it contained a universal norm bound by a homogeneity marked as "women" instead of "nation." This particular unraced protocol feminism can further be related to what historian Doug Rossinow calls a "postscarcity activism" associated with a host of radical politics in the postwar period, such as the student New Left Movement. Rossinow defines *postscarcity activism* as a kind of politics that emerged in the affluence of the post–Second World War years and oriented itself toward extending privileges and finding authenticity, rather than ameliorating immediate deprivation.[46] This type of activism contrasts with, for example, the urgency of the civil rights movement relative to the deprivations and violence authorized by segregationist states during the same period.

In this sense, feminist self help was not about survival but about the possibility of "taking back turf" and "equalitarian" practices. Doing so meant maintaining maximum health within a historical condition of relative affluence that made claiming sovereignty over oneself seem available. This feminism's democratic protocols that emphasized individualized possession of the body aspired to avoid racism, sexism, homophobia, capitalism, and other sources of oppression by putting the singular woman in charge of the clinical encounter. The achievements of feminist protocols in successfully individualizing choice in clinical moments underwrote a bracketing off—an avoidance of explicitly reckoning with—the many "interlocking" or "intertwingled" systems converging on living-being. At the same time, feminist self help called upon women to make themselves available to, rather than avoid, technical acts on their bodies in the name of a transformed politics of wellness and sisterhood.

By contrast, the writings of the Combahee River Collective and affiliated women of color feminists of the 1970s offer a very different description of the entanglement of health and subjectivity in the aftermath of the civil rights movements.[47] Politicizing health could not be isolated to the scope of clinical protocols and instead was inseparable from other aspects of racialized political economy—labor, citizenship, welfare, housing, incarceration, sexuality, and policing. Importantly, questions concerning birth control, abortion, and sterilization within much black feminist writing of the 1970s were figured as caught within *contradictory* forms of devaluation, for example, between histories of slavery, contemporary policing practices, and devaluations of black "matriarchy" rampant at the time.[48] In other words, how people were differentially made available to harm, via race and capitalism, was as central to their health politics as the ways people were called to foster healthfulness and claim self-sovereignty. Racialized violence was materially distributed and lived, but did not exhaust the meaning of lives. In contrast to the figure of the already "healthy" woman in unraced feminist self help, critical writing by black feminists repeatedly characterized the collective biopolitical condition of black women using the words of the civil rights activist Fannie Lou Hammer, etched as her epitaph: "sick and tired of being sick and tired."[49]

Uninvestigated murders, police violence, continued segregation, and incarceration formed a material biopolitical context in Boston that the Combahee River Collective attempted to portray and intervene in. They charged themselves with contesting "how little value has been placed

1.3. Third World Women Alliance banner in front of community organizer Marlene Stephen, as she speaks at a Boston march in 1979 concerning the failure of the police to address recent murders of young black women. Images of the banner and its slogan were used widely in community-organizing posters and in the Combahee River Collective's pamphlet "Why Did They Die?" The Third World Women Alliance was strongly affiliated with black feminism and had origins in the Student Non-Violent Coordinating Committee. Photo by Ellen Shub, courtesy of Valerie Stephens.

upon our lives." In addition to their "A Black Feminist Statement," the Combahee River Collective published a pamphlet called "Six Murdered Girls: Why Did They Die," analyzing the interacting racism and sexism that underwrote police complacency about a series of murders in 1979 (eventually totaling thirteen) in black Boston neighborhoods, helping to rally community protest. At the resulting demonstration, a prominent banner read: "3RD WORLD WOMEN: We CANNOT LIVE WITHOUT OUR LIVES" (see figure 1.3)[50] Rather than announcing the liberal sovereign subject of unraced feminism, the banner's phrasing announced what could be considered the politicized subject of *necropolitics* (the governing of death, its distributions, forms, and likelihoods) brought to life and the marked imagined community of "third world women."

In other words, the Combahee River Collective was exemplary for the way it attended to and sought to subvert the necropolitical, death-producing, effects of biopolitics in the late twentieth century. Distributions of vulnerability to state violence; availability to exploitation; precariousness relative to denials of full citizenship were conditions as central to the biopolitical topology at this time in the United States as were injunctions to live as self-possessed healthful subjects. Scholars as diverse as Achille Mbembe, Giorgio Agamben, Dorothy Roberts, Sarah Lochlann Jain, Andrea Smith, and Melissa Wright have insisted on this attention to the necropolitical dimension of biopolitics — the selective fostering of life *and* death. Biopolitical formations do not only foster living-being as a site of efficiency, labor, sovereignty, value, safety, and so on; they also designate zones of "lives less worth living," less valued, more available to neglect, injury, precariousness, abjection, and open to violence not conventionally counted as such. Necropolitics, again, can be understood doubly, at once analytically as the material distributions of death and vulnerability to violence inherent in practices that take life as their concern, as well as the historically specific politicization and naming of these distributions as kinds of politicized death.[51] While the material expression of biopolitical topologies can be understand as always also necropolitical — that is, composed of uneven death worlds as much as injunctions to live — historical actors have selectively recognized and politicized this dimension in their projects to politicize life. Importantly, whiteness — in heralding self-possession — often works to displace attention from the necropolitical work of race.

A politicized subject-figure that lives at the contradictions between calls to life and availability to death — a kind of living dead, or life despite of and in resistance to death, recurs in much black feminist writing about health in the 1970s and 1980s. Byllye Avery, a founding member of the Gainesville Women's Health Clinic, described the condition of black women in the strong terms of "dead inside" and "walking around dead."[52] In Los Angeles, Mother's Anonymous, a group within the Aid to Needy Children (ANC) organization (discussed later in the chapter), explained their use of the word "anonymous" for its meaning as *nameless*: "we understood what people thought about welfare recipients and women on welfare was that they had no rights, they didn't exist, they were a statistic and not a human being."[53] Reproduction was likewise situated in this ambit of necropolitics. For example, accusations within black nationalism that birth control contributed to black genocide were rebutted with the manifesto "The Sisters Reply" written in the 1970s by the black radical feminist New York Mt. Vernon Group. They reversed this gradient of death effects, by arguing "for us, birth control is freedom to fight the genocide of black women and children."[54] In contrast to the unraced "woman" of LA feminist self help, this raced and sexed figure of the "living dead" claimed self-determinism from within collective and contradictory conditions of dispossession.

Taking the Combahee River Collective as an example, one can see that the politicized subject of feminism was fashioned in the fractures of contradictory dispossessions in ways both affective and binding. On an affective register, the contradictions of dispossession were often described as crazy-making and painful, politically needing to be "made sense of" as a joint matter of psychic survival and political tactic of coalition.[55] The subject position portrayed in the "Combahee River Collective Statement," for example, was not one of singular identity but of a fragmented "identity politics" informed by a series of disidentifications and irreducible frictions—with heteronormativity, with whiteness, with nationalism, with capitalism—that necessitated multiple and partial allegiances. Health, in this sense, was a site in which dispossession and injunctions to foster empathy converged and clashed, and thus were not adequately remediable with the tactic of reappropriating medicine. Instead, protocols for "taking control of reproduction" were a partial, though necessary, oppositional tactic in creating conditions for livable lives.

Helen Rodriguez-Trias, a Latina feminist doctor and founding member of the Committee to End Sterilization Abuse, aptly described white women as focusing their energies on *health care* (service provision and autonomy in care) while women of color and working-class women emphasized *health status* (the structural conditions that determined health).[56] While that distinction may not hold for the example of projects such as the working-class Somerville Women's Health Clinic, divergent articulations of privilege, dispossession, service, and suffering provide compass points for this chapter's effort to map the work of race and the various politicizations of "health" among different feminisms.

Importantly, an ethos of starting with oneself traveled across and bounded these various articulations of feminist health politics. Works by the Combahee River Collective, Audre Lorde, Cherrie Moraga, Barbara Smith, and many others insisted on taking *one's own life* as an epistemological and affective starting point from which to chart the violence and possibility of a contradictory multifaceted politics of living. Similarly, within feminist self help, protocols invoked self as both the starting place of politics and health care, and the target of action—you did self help to yourself. Medicine did things to you; feminists took charge of themselves. Holding together the Combahee River Collective and unraced feminist self help in the same historical frame marks how these projects were both animated out of the uneven racial formations composing Cold War America, and hence generated entangled yet divergent diagnoses of the 1970s.

It is on this larger, uneven biopolitical terrain—through this double vision that understands feminism as both symptom and diagnosis of its moment—that I want to place the founding narrative of feminist self help. This narrative, far from highlighting connections across simultaneous yet discrepant feminisms, situates feminist self help in a moment of spontaneous generation on the evening of April 7, 1971. On that particular evening, so the narrative goes, a score of frustrated feminist abortion activists gathered in the front room of a Venice Beach house that served as a feminist bookstore, gathering point, and home of the *Everywoman's Newsletter*.[57] The topic for discussion was the possibility of learning how to perform abortions themselves, along the lines of the underground abortion service in Chicago known as "Jane."[58] Sitting in a circle on the floor, those gathered, mostly white, took turns introducing themselves and describing the scope of their political work—protests and abortion referrals—with an air of dissatisfaction.

Carol Downer was one of the organizers of this meeting. When her turn came around, Downer demonstrated the method of vaginal self-exam. Lorraine Rothman, who with Downer presented herself as a founder of the feminist self help movement, narrates her recollection of this first meeting vividly:

> She takes us into the adjoining room and pushes everything off the desk, and then goes around and pulls down the shades—I mean this was an old house—pulls down the shades in each of the rooms. Goes to the backdoor, locks the backdoor. Goes to the front door, locks that. And I'm thinking, oh my god! Woo, what did I get myself into?! And all the time she's talking, very quietly. . . . slowly and very clearly. . . . and she's talking about how she was so impressed with this *thing*. And while she's talking she removes her underpants, puts them aside, and she gets up on the table—she doesn't look at anybody's face—and gets up on the table, positions a pillow she had already prepared, and pulls

up her skirt. She had a very long flowing skirt that she could control to drape over her legs. And she shows us this plastic vaginal speculum, which I had never seen before. I'd never seen a speculum before and yet I had umpteen, umpteen, umpteen visits. . . . and I've had kids! . . . And she says, "what I'm going to show you . . . ," and she goes through this whole process, and inserts it into her vagina. She says she's going to be really careful. It's not uncomfortable, and since she's doing it to herself, she can control it herself. And since the handle is up, she can open it as wide or as little as she wants. And she proceeds to manipulate it. She uses a flashlight and mirror to project for herself and make sure her cervix is in view. . . . And then she says, "Would you like to see?"

We were all standing there all around her absolutely, totally amazed at what she was able to do. And the fact that this particular area of the body that has been inaccessible to us is now *visualized*. . . . It was so revolutionary! Just the simple act of putting a speculum in the vagina ourselves and bringing up that part of our body and being able to see it in the same commonsense way we look at our face every morning.[59]

This oft-repeated founding narrative offered a particular portrait of feminist self help politics, of which three elements are worth emphasizing.

First, it portrayed a "world turned upside down," a moment of revolutionary innovation and spontaneous generation presented as enlightenment or a "click" — of seeing newly and clearly what had previously been obscure — and of liberation, of offering a route out of repressive reproduction. Though many of the women present were abortion activists, most did not themselves know how to perform an abortion, even if they had gone through it themselves. It was quite probable that some attendees had never before looked inside their own vaginas, and the literature from this era is replete with testimonials of the "consciousness-raising" effect of the vaginal self-exam's repositioning of women as both observer and observed, patient and practitioner. Self help presentations were often testified to produce a "click" — a sound from popular photography of the era.

The historian Brian Beaton has noted the mechanical referent in this oft-cited effect of consciousness raising — the "click" sound of a sudden technical ratcheting into place or of a camera producing a picture.[60] The click might be read as the snapping sound of subjects viscerally placed in new conjunctures, of subjectivities produced in new positions, with new ways of seeing the world. In general, the "click" tended to occur for women whose social locations were legitimated by racialized ideologies of bodily respectability and modesty, who could reference their bodies as "healthy," who experienced sex as the primary site for the exploitation of their living-being, and less so for women whose bodies were made regularly vulnerable by other topologies of exploitation. In other words, the political traction of *operability* differed across strata of biomedicalization and racialization. Injunctions to constantly monitor one's body, to turn oneself into an experimental subject, to already be normal, to open one's bodies to the view of others for the sake of an aggregated group, "clicked" in particular ways for lives already valued through injunctions to be regular and compliant patient-consumers, injunctions shaped by whiteness, biomedicalization, and emergent acts of neoliberal subjectivity. In the intensely racialized conditions of the United States in the 1970s, the attraction to the click of "operability" could conspire to undermine a precarious corporeal integrity.[61] At the same time, feminist self help offered a radical overturning in how authority in health and biomedicine was habitually understood.

Second, this founding narrative underscored practice—the demonstration and how-to, the sharing of protocols. Like other radical feminists learning how to repair cars or build houses, feminists self helpers were appropriating the skills of a male-dominated trade. Constituted through protocols, feminist self help insisted, why ask permission for something you can do yourself?[62] The simplicity of vaginal self-exam helped to exemplify this tactic of reappropriation. Not only was it an easy noninvasive procedure, it was claimed to reveal a simplicity to the body itself: "We learned, hey, the cervix is just a couple of inches in there, it's not all curlicues, and caverns, and passageways."[63] Further, the ease of vaginal self-exam suggested the possibility of learning and hence appropriating other techniques, such as abortion. It taught, according to Downer, "how easy it was to learn these things—that they were learnable. They were not rocket science."[64] With vaginal self-exam as an inroad, feminist self help articulated a political program of "taking back turf" from medical authorities by learning and reassembling the techniques of "reproductive control" by experimenting directly on one's own body.[65]

Lastly, this founding narrative emphasized the "revolutionary" politics of these practices by foregrounding the need for secrecy—shades drawn, door locked—as well as the inspiration of black-market, underground health services. In this way, the founding narrative made clear that these practices were a form of radical counter-conduct, an escape from the dominant, a tactic of liberation.

Announcements of spontaneous enlightenment, technical appropriation, and revolution, as this founding narrative recites, can be reread as a tactical assemblage of a feminist health politics that was at the same time a narrowing of the ambit of politicization—not interlocking oppressions, but the scope of health care practices. The narrative offers a lens into both the excitement of finding politics in the nitty-gritty of technoscientific practice, and the legible production of a new conjuncture between reproduction, the aspirations of liberal subjectivity (unraced, yet sexed), and the promises of technoscience at this moment. Described as a mean to liberate reproduction from "power," feminist self help protocols were understood as acts of "taking power."

I want to take this taking power seriously, both in terms of the historicizable way power was often theorized in radical feminism as something which could be possessed and appropriated, and as the sign of a shift in the liberal governmentality of health beyond the state, formal medicine, and professionalism into the sites of the nonprofit organization, the participatory small group, and the responsibilized individual. "Not rocket science," reproductive health was reassembled as accessible, routinizable, actionable, appropriable, and rearrangeable. As a counter-conduct, feminist self help sought to tactically "seize" power, as portrayed in the well-known cartoon, given to Downer, of a scantily clad Wonder Woman snatching a speculum from a doctor's hand and wielding it against the cowering figures of the pope, the district attorney, and the American Medical Association (see figure 1.4).[66] At the same time, Wonder Woman, while ironically depicted, was also a Cold War heroine of sexualized white womanhood intended by her creators as a prototype of a new liberated American feminist subject—the red, white, and blue amazon queen of the world's empowered women. Recast as an individualized version of feminist struggle who could rise above and seize the tools of her oppressors, she graced the first issue of *Ms. Magazine*.

Feminist self help, then, was expressive of a larger feminist practice of imbuing subjectivity as a starting point for articulating politics, but mobilized as a particular version of this politicized subject: American, unraced, unclassed, yet sexed, and individualized as the starting point for a flexible ethos of self-governing. Thus, feminist self help as a biopolitical project was both a strategy of counter-conduct that sought to "take control" of

1.4. The well-known poster of a speculum-wielding Wonder Woman, given to Carol Downer when she was acquitted of the charge of practicing medicine without a license in December 1972. The poster graced the walls of the Los Angeles Feminist Women's Health Center. Courtesy of Lorraine Rothman.

one's own living-being, as well as a symptom of the fractures a feminist self-sovereign subject produced, for whom sex alone was elevated as a privileged diagnostic lens.

Biomedicalization and the Politics of Health

Feminist self help was, thus, a historically particular way of foregrounding subjectivity in efforts to designate life and protocol—bios and techne—as domains of politicization and reassemblage.[67] Yet, feminist self help did not spontaneously invent this bind between life and techne in 1971 Los Angeles; instead, this bind was part of larger investments in the politics of "health." Health as a political concern was not new in 1970s but was brought into new vital forms of value and politicization. Notably, health became institutionalized as an international human "right" in the World

Health Organization's Alma Alta Declaration (1978). Ethicized as a universal right, health was also increasingly described as an indeterminate and extensive form—more than simply the absence of disease, it was a more general state of well-being at physical and mental registers.[68] *Health* in this sense was reinvigorated in the 1970s as a domain of value that authorized both national and transnational interventions. Indeterminate in form, "health" underwrote an array of projects, from community clinics, to environmental regulation, to international development regimes. In other words, the qualities of indeterminacy and extension were productive of intensive investments into health as a domain in which questions ranging from decolonization, to increased GDP, to women's liberation—and not just illness—were at stake.

Drawing on Foucault's methods to historicize "race," the historian Ann Laura Stoler describes scholars' contemporary commitment to insisting that race is constructed and flexible (rather than biologically fixed) as not simply an antiracist antidote to past scientific racisms. The commitment to underlining the constructed and changeable quality of race is also a symptom of our current "regime of truth" in which the indeterminacy of race is what authorizes its continued circulation today.[69] Similarly, one might argue that health as an expansive and indeterminate domain at stake in virtually all human activity can also be thought of as a historical regime of truth in which health authorizes and ethicizes a diverse array of governmentalities with diverse material effects.

This valorization of health as a capacious ethical domain, moreover, was accompanied not only by a proliferation of projects in its name, but also the intensification of health as an economic enterprise. Thus, in the United States, the expansion of "health rights" through such state projects as Medicare and Medicaid was at the same time an invigoration of "health systems" requiring new services and commodities—from insurance, to drugs, to medical supplies, to data management—forming a growing economic sector, dubbed the "medical-industrial-complex" by its critics.[70] Health, was retheorized by economists as contributing to a nation's "human capital" (understood as embodied capacities to be economically productive) at the same time that individual aspirations to well-being could be harnessed to consumer acts.[71] In the 1970s, the not-yet-sick, at-risk patient, emerged as a central figure of medical intervention, as predictive laboratory-generated criteria could be mobilized to identify risks as virtual pathologies (such as high cholesterol) and create new drug markets to treat possible pathologies preemptively. The emergence of the not-yet-sick, at-risk patient joined physicians' efforts to foster well-being in preemptive and preventative ways with drug companies' investments in diagnoses as a profit-generating practice in a teleology the historian Jeremy Greene calls "prescription by numbers."[72]

In other words, the 1970s helped to inaugurate some of the infrastructural elements of a complex and multilayered reconfiguration of health care named "stratified biomedicalization" by the feminist technoscience scholar Adele Clarke and her collaborators.[73] The emerging features of stratified biomedicalization, in turn, sparked new sites and contradictions for politicization. For example, the extension of the U.S. state into health care through Medicaid and Medicare was at the same time an enfranchisement in the name of "health rights" responding to the medical civil rights movement, *and* a reconfiguring of health care services that distinguished between those citizens who had to interface with the state to achieve health care and those who could afford to circumvent the state, and hence both its injunctions and negligence.[74] Moreover, biomedicalization featured both the elaboration of medicine as a corporate for-profit endeavor and a multivocal appeal to "responsibilize" individuals as accountable for their own health. It was typical of the contradictions within stratified biomedicalization that calls for "basic" or "primary" health as a human right, which brought mass constituencies into medical care, occurred in the same moment of proliferating cultural practices of "healthism," which in turn was amplified by the ways consumption and identity politics fostered individualized ethical practices of self-care and self-improvement.[75]

Protocols were at the center of biomedical logics, expressed in data and management "systems" that governed risk and rationalized clinical and administrative practices in ways that tied together health, efficiency, and profit. Clinical encounters became not only moments when commodities could be dispensed, but were also generative of research data that could then be further used to tailor treatments.[76] Data collection — computing, record management standards, and databasing systems — facilitated not only the clinic/research circuit but also epidemiological and probabilistic calculations instructing individuals how to reduce risk — to not smoke, to wear a seatbelt, to have a Pap smear. Moreover, doctors as agents within these systems were themselves increasingly subject to standardized protocols, reducing personal professional authority, standardizing reimbursable diagnoses, and circumscribing the scope for judgment.

Feminist self help emerged, then, precisely at this dawn of biomedicalization, when protocols were invented and proliferated as organizational efficiencies and individualized responsibilities.

Reproduction—as a feature of capacious and indeterminate health—likewise was a domain of stratified investment, with the 1970s marking the rise of boutique preventative gynecology for middle-class patients as much as the eruption of state-sponsored family-planning programs directed at poor and racialized communities. Family-planning projects, in turn, harnessed individual desires to manage fertility to state economic ambitions and racialized class politics.[77] Limiting births among the poor as a form of national and postcolonial economic development, border control, and poverty eradication reached its high-water mark in the 1970s, with the United States as the single-greatest investor.[78] To self-govern one's own fertility with medical commodities became not only a purported sign of the arrival of a female modern subject who chooses in a cost-benefit matrix, but also a contribution to national and global well-being.[79] While scholars have tended to locate this entanglement of family planning and economic development as a core feature of postcolonial sites, particularly in Asia, as well as a feature of coercive communist state regimes within Cold War geographies, this history also touches down inside the United States.

From the mid-1960s onward, under the rubric of "family planning" instead of "population control," the promotion and distribution in the United States of federal and state-subsidized services spread through doctors' offices, hospitals, emergency rooms, housing projects, employment projects, and specialty clinics.[80] Unlike some postcolonial states, no single U.S. governmental department was charged with the project of family planning. Instead, family planning was a filigree formation, at stake in a crazy quilt of projects and sites. Without an overarching state department of family planning, subsidized services were often channeled through nonprofit organizations, themselves hitched to a new array of practices and procedures. Not without resistance, and in disturbing contiguity with eugenics legislation already on the books of state legislatures, fertility became a stratified zone of investment within the capacious domain of health.

Thus, this period saw an explosion of guidelines, protocols, and handbooks concerned with family planning, often coordinated by such extra-state organizations as Planned Parenthood, the Guttmacher Institute, or the Population Council, creating circuits of connection between domestic family planning and its transnational itineraries. These circuits were further supported by a "gold rush" of cheap mass-produced hormonal pharmaceuticals, a new generation of plastic contraceptive devices, and federally funded surgical sterilization in public hospitals.[81] In other words, feminist self help's focus on protocols was crafted in a particular moment when "health"—and fertility as a correlate of "health"—was actively reanimated as a racialized and sexed concern of emergent rearranging postcolonial and Cold War governmentality, altering the "inside" of the United States as much as the "over there" of the population control programs the state supported.

Los Angeles was one city among many where this complex emergence of family planning expressed itself. Jokingly called the "capital of the third world" because of the many diasporas that converged there, LA had one of the densest concentrations of family-planning services in the country.[82] Between 1967 to 1971, forty different clinics focusing on family planning were established there, sparking a municipal council to coordinate these scattered efforts, standardize record keeping and training, centralize data gathering, and create uniform "protocols."[83] Family-planning practices were not just performed in designated clinics, but also in doctor's offices and public hospitals. A 1971–72 survey found private doctors in Los Angeles more likely than the national average to recommend and offer family planning to patients.[84] Practices in public hospitals reflected national attitudes about the reproductive fates of poor women: a 1972 national survey by Planned Parenthood asked doctors to consider five possible actions for an unmarried woman who just had her third out-of-wedlock child: 40 percent said they would refer her to psychological care, 96 percent would inform her about sterilization, 36 percent would support removing any future children from public assistance, and 30 percent said she should stop receiving governmental welfare unless sterilized.[85] Los Angeles County Hospital, in turn, was particularly notorious for coercing patients of color to get sterilized during labor.[86] Thus, while feminist self help was certainly a vociferous critique of past oppressive medical practices, it was also a reaction to current developments, in which reproduction was a site of renewed and intensified intervention by society, which had created new methods to bind individual desires with medicine, race, and national economic futures.

It was in response to these stratified conditions that health in general became a domain of counter-conduct in the 1970s. In Southern California alone, the varieties of counter-politicization of health included the Los Angeles Free Clinic, which operated out of a small store front; the Bunchy Carter Free Medical Clinic in South Central LA, as part of the Black Panthers' "survival pending revolution" strategy; and the United Farm Workers' health clinic for migrant agricultural workers, which operated at the site of the famous Delano strike.[87] Even from within the medical profession, counter-conduct projects occurred. The University of California, Los Angeles, was the site of a medical student social movement, as well as where radical nurses fostered interventions and analyses of the provision of community health and family-planning services to "people of color."[88] At a national level, the Johnson administration's "war on poverty" had funded community health centers scattered across the country, which were often subverted as a politicized form of community-run service.[89] For many radical health projects, the politicization of medicine existed in a larger transnational circuit of anticolonial politics—as expressed by the psychologist Franz Fanon and the physician Che Guevara, or even for some in Maoist health policy.[90] Thus, health was politicized as both a site where dispossession expressed itself in bodies and as where liberation could be won and remediation achieved. Within Los Angeles, such sentiments found more conventional expression in a gubernatorial report on the causes of the 1965 Watts riot. The report attributed lack of health care as a primary reason for the riots, leading to the construction of a public hospital as one of the few concrete attempts at revitalizing the neighborhood.[91]

As with Boston, in Los Angeles the feminist forms of this efflorescence of counter-politicization of health were polyvalent. Within Los Angeles, in the same moment that feminist self help crystallized, racialized groups took action in different ways. ANC Mothers Anonymous was formed by black women in Watts as one of the first grassroots organizations of mothers receiving welfare, Students in Long Beach established a Chicana *feminista* group and newspaper, while Los Angeles Chicanas filed a law suit against the Los Angeles County Hospital with charges of coercive sterilization. Asian Pacific health activists conducted a survey of health services for Asian women, which grew into the Asian Women's Health Project, as well as founded health clinics, such as the To Help Everyone (THE) Clinic for Women in Santa Barbara, organized as a service for women of color.[92] Yet another Asian feminist group from UCLA organized a service for youth drug users.[93] Health was thus a contested site for a multiplicity of feminists who generated myriad expressions of counter-conduct, often oriented around racialized communities, necropolitics, and service provision. In short, in Los Angeles of the 1970s, as much as in Boston, health politics was a problem-space produced through the contradictory convergence of violence and coercion against hegemonically devalued lives as much as through injunctions to live more abundantly and freely.

Within these manifold politicizations over health, feminist self help was a particular kind of biopolitical project that sought to "take over" not just health care acts, but participants' own "lives" and "bodies" by virtue of creating protocols that situated reproduction as the technical and ethical responsibility of individual women over themselves. While this ethic of individual responsibility at times echoed the rising individualization of health care within an incipient neoliberal logic, it was nonetheless also a particular critical diagnosis that declared the sexed individual as only politicized within a larger collectivity, never alone. Feminist self help appropriated biomedical protocols, attempting to craft techniques that would disengage sexed bodies from medical and state investments into reproduction and instead charge responsibility over sexed life to individual women themselves. Feminist self help sought to disrupt the simple equation of a liberal self-sovereign subject with practices of individualized consumption by insisting that individual acts alone were not feminist; they required a collectivity of women to become political. But how to assemble this collectivity?

Social Technologies and the Politics of Small Groups

"A self help clinic is not organized to serve the community of women — we are the community of women."[94] Self help was a recursive practice: it called on a "community of women" who then sought to individually determine their own bodies, a task which could only be performed in a group of women. As something you do to yourself and not to others, feminist self help explicitly tethered the ingredients of the self-determined health practices to the unit of the "small group": a set of participants who, ideally, would be "sisters." The small group was thus the social platform that translated individual subjectivity and technical acts into the ambit of feminist biopolitics (see figure 1.5). Formed through protocols of sociality and discourse — that is, guidelines for how to interact and talk with

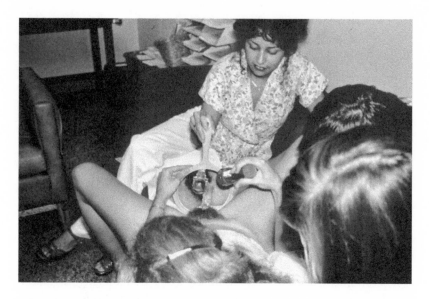

1.5. Photographs like this one emphasized the small group format that turned vaginal self-exam into a collective act. The photograph positions the viewer in the gaze of the woman under examination. This photo was a slide used in the Los Angeles Feminist Women's Health Center presentations of the 1970s. Courtesy of Lorraine Rothman.

one another—small groups were used to both create political "consciousness" in participants, and hail participants as "change agents." Within the group, protocols also directed the affective entanglements between members, who aspired to create antiauthoritarian and supportive conditions in a microcosm that would make self-sovereignty in caring for oneself possible. In other words, the small group was a politicized terrain, where protocols of sociality evoked what Paul Rabinow has called "biosocialities," in which social collectivities are knit around identities formed through the possibilities for technically intervening in living-being.[95]

Small group work, in the form of consciousness raising, was pervasive within feminism, though the exact methodologies and results varied widely.[96] While the feminist self help group was certainly part of that history of feminist practice, its specific protocols for small group work deserve closer scrutiny. While the next chapter looks more carefully at the epistemological work of consciousness raising as a research protocol, here I want to track discrepant genealogical threads that helped to elevate the small group as the core unit of feminist self help: a sociality, moreover, politicized and governed through protocols. The small group of feminist self help was a platform for the hailing of individuals as politically and historically important actors, and in so doing was part of a wider history of techniques of group work circulating in Cold War democratic, radical, and emerging neoliberal governmentalities.

Early radical feminist consciousness-raising advocates of the late 1960s were directly inspired by the participatory democracy methods of the Student Non-Violent Coordinating Committee (SNCC), including the slogan "let the people decide," which captured a belief that radical movements were not directed at people; they belonged to the people.[97] In turn, the SNCC's methods were themselves indebted to broader practices of fostering "black consciousness" within transatlantic black radicalism that included Garveyism in the early twentieth century and the Black Power movement in the United States, but also the South African Black Consciousness movement and the ideological framings of Steve Biko.[98] Here the question of consciousness was an aspect of "decolonizing the mind," drawing on the psychological and existential turns in both Fanon and Albert Memmi.[99] Similarly, the practice of "speaking pains to recall pains" in the Chinese Revolution, as well as Mao's injunction to trust your "physical sense organs" over books, served as oft-cited models for the development of consciousness raising at a moment when reading the "Little Red Book" in the United States was a badge of radicalism.[100] While radical feminists, unlike socialist feminists, tended to highlight patriarchy over capitalism as the structural source of sex oppression, they were nonetheless deeply critical of capitalism as a formation of patriarchy and were clearly genealogically indebted to Marxism and its practices of self-criticism.[101] Further, feminists' emphasis on coupling a search for personal truth with "action" reverberated with the existentialist and Christian liberation theology threads of the American New Left in the 1960s.[102]

While feminist self help group work was crafted amidst this tangle of related radical and Marxist practice, it was also symptomatic of a broader "psychologization of politics" in the United States.[103] Radicals were not the only ones practicing small group work, sitting in a circle, using antiauthoritarian "facilitators," and stressing the "cooperative" examination of oneself. Los Angeles was an epicenter of small group work in the professional field of human relations, which in the 1970s took the forms of "encounter groups" and "sensitivity training" practiced in areas as diverse as business management, medical training, schools, and group therapy. This more conventional lineage was not only geographically proximate to

Los Angeles feminist self help; it is also present in many of the words used within their manuals and pamphlets, such as the unusual term *equalitarian* and the phrase "change agent."

The historian Laura Kim Lee has excavated a genealogy of such "encounter groups," tracing their techniques back to post–Second World War "human relations" science in the United States.[104] Human relations research was concerned with investigating the terms of "equalitarian" relations and "group dynamics" so as to encourage democracy and manage "social change." In many ways feminist self help, as a protocol feminism, joined the goals of psychic decolonization—phrased as "demystifying" and "reclaiming"—together with the technical protocols from Cold War human relations methods, for which Los Angeles was an important disciplinary location.

The Jewish German psychologist Kurt Lewin was a central figure in the establishment of human relations research, immigrating to the United States with the rise of German fascism. His work set out to experimentally derive the difference between "autocratic" and "democratic," as well as "laissez faire" social behaviors among children. He argued that groups had interpersonal "dynamics" that could be scientifically uncovered, in turn leading to "social technologies" that would foster more democratic group relations.[105] In his experiments, small "groups" were the objects of study, but also the instrumentalized "tools" through which laws governing interpersonal relations could be drawn out.

While the term *group* may at first glance appear neutral, in the Cold War period the fear of fascism and communism invested the unit with a set of overlapping anxieties. "Modernity"—with its large corporations, sprawling bureaucracies, institutionalized schools, and mass culture—purportedly threatened the "individuality" necessary for democracy. Large groups were thus dangerous units in which authoritarianism might fester.[106] At the same time, within human relations research, smaller "groups" were seen as the crucial unit in which the development of democracy and autonomous individuality were at stake. Lewin's work was itself preoccupied with this sense of both the danger and promise of groups. Rather than just developing the techniques of democracy, he consulted for the Office of Strategic Services (the precursor to the CIA).[107]

By the 1950s, human relations research understood "groups" both as an experimental technology for their research and as a "social technology" or a "device" that could produce "social change" in daily life within the context of the Cold War. Rather than primarily as a political theory of government, democracy was considered by Lewin and many other human relations researchers as a set of behaviors and relationships that could be learned and practiced in everyday lives within groups. Democracy was thus defined as a social technology of the group.[108]

The opening of the National Training Laboratory for Group Development (NTL) in Bethel, Maine (initially funded by the U.S. Navy and the National Education Association) signaled two of the disparate domains in which human relations research would flourish: state agencies and education. Through "training groups," or "T Groups," research subjects—called "delegates"—could be taught to become aware of interpersonal dynamics, in the process transforming participants into "change agents" that would encourage democratic rather than "autocratic," or later "totalitarian," ways of behaving in organizational structures.

Within the NTL, T Groups were a laboratory method, a "device where, in an initially unstructured setting with the usual group controls absent, the members develop group norms, standards, power and friendship structures, patterns of communication, and shared problems on which to work. In the process they analyze their own behavior and that of others in the group, sharing these observations with other members to gain both personal skills and insight, and knowledge of group function."[109] While at first researchers held themselves apart from such groups, they later added "feedback" practices in which the delegates took an active part in reflecting on and critiquing the experiment, and even were encouraged to innovate through their own behavior. Feedback was seen as a crucial innovation of this research, having a powerful "electric effect" on participants and staff.[110] With feedback, researchers became "facilitators" who joined in the group and modeled desirable group techniques, evoking a helper rather than an expert relationship. The NTL focused on training school teachers and social workers (many of them women), but also corporate executives and state bureaucrats, as "change agents" who would learn to foster democratic forms of "cooperative group process" rather than authoritarian productivity in schools, bureaucracies, and businesses. Even the organizational structure of the U.S. Department of State under Kennedy was revised with the help of just such human relations consultants. In turn, Kennedy's "self help" foreign aid doctrine further dovetailed with the spread of human relations techniques into foreign aid programs.[111]

Southern California soon became an influential nexus for human relations research. In 1954 the Western Training Laboratory opened at UCLA. Closely associated with the business school, its particular focus became "sensitivity training" for executives and managers.[112] In its Californian iteration, the practices of human relations research shifted away from its concerns with democratic change agents toward helping participants reach their "human potential," emphasizing the emotional component of group dynamics. In this shift, human relations research became entwined with "humanistic psychology" that emphasized "self-actualization" as a human need and goal, further accentuating the personal therapeutic and revelatory effect on participants. In La Jolla, California, a leading figure in humanistic psychology, Carl Rogers, started the Western Behavioral Science Institute, which transformed the T Group into the "encounter group," a formulation loosened from its instrumentality in research to fully become a method of "self realization" or "self actualization."[113] Rogers helped to popularize encounter group methods in his writings and through the use of films, including the Oscar-winning documentary *Journey into Self.*[114]

Humanistic psychology, in turn, mapped a five-staged pyramidal "hierarchy" of human needs, starting with the "basic" physiological needs of eating and drinking (or what could also be called bare life), moving up to the need for safety and security, to social needs of belonging and love, to self-esteem, and finally at the top to self-actualization (what also could be called the self-sovereign subject).[115] The pyramid can be read, on the one hand, as a particular diagnosis of the uneven biopolitical arrangement of American life—ranked from bare life to self-determination—and on the other hand, as the relative revaluing of group work through an emergent liberal governmentality that valorized individualized empowerment and productivity within social groups. Emphasizing emotions and the uniqueness of each individual, as well as the value of sensual and social awareness, encounter groups and sensitivity training in the 1970s could be found in California from medical schools, to elementary schools, to the department of state, to the radical feminist techniques of sitting in a circle as "group process."[116]

In short, in the 1970s altering individual consciousness with the "social technology" of the small group was a circulating protocol within what were then incipient neoliberal practices, as much as it was a radical method in projects of counter-conduct. It is possible to chart the rise of a new liberal expression of subjectivity articulated through the disseminated protocols of human relations research, protocols composed through techniques for examining, valuing, and naming one's self as a form of self-actualization. In turn, self-actualization as a late twentieth-century art of governmentality, primed the further unfolding of what the historian Ellen Herman has called the "psychologization of politics" and what the social theorist Barbara Cruikshank has called the normative "will to empowerment."[117] The project of knowing and fulfilling individual desires could easily slip into practices of consumption and marketing. Or it could fit into managerial regimes that sought to harness emotive relations and creativity as forms of "human capital," another economic term emerging in this period.[118] As the service and "knowledge" sectors of the U.S. economy expanded, accompanied by changing racializations and genderings of the workforce, *affect* (as emotion, capacities to empathize, capacities to learn, capacities to be creative, capacities to interrelate, and so on) became the explicit focus of a newly psychologized managerial profession, from the encounter group in the 1970s into diversity training and the emphasis on managerial "soft" skills in the 1980s.[119]

Feminist self help, as a project of counter-conduct within the "small group," was animated amid this extension of emotionally charged small group practices into so many facets of everyday life, even as it fostered a collective, noncapitalist project. From its language of "facilitators," "antiauthoritarianism," and "equalitarianism," to its emphasis on supporting unique individuality in a cooperative group setting, to its insistence on crafting learnable, portable, and transmissible protocols for "group process," feminist self help was at least partially appropriating human relations techniques, as well as participating in the broader elevation of small group labor. Today, feminist consciousness raising—not human relations research—is celebrated for fostering social change through small group work. While feminists typically took pains to differentiate consciousness raising—intended to result in political action—from group therapy, and while these distinctions are important (and explored more fully in the next chapter), the two sets of practices were nonetheless entangled in the shared elevation of the small group as a technology of social change.

Feminist self help was, ideally, rather than a service for others, a practice done for "ourselves." The "ourselves" could be the members of an intimate small group, but in a more formal feminist clinic it could also be capaciously invoked as the "community of women." In feminist clinics, aspirations to "structure out authoritarianism" with "women-controlled" collective organizations fiercely clashed with the larger logics of hostile state licensing boards and labor relations. Who wrote manifestos, who did promotional road shows, and who cleaned up? Downer tried to differentiate between the labor politics of two kinds of feminist projects: one, feminist women's health clinics as a kind of social service work, prone to exploiting good intentions in the name of an overarching goal, and the other feminist self help as a project of appropriating one's labor for oneself with the help of others.[120] Feminist self help, then, was fostered as a site where the emotional and social labor of collective work, health care, and detailed self-observation aspired to escape from circuits of profit and law, avoiding the problem of labor in microcosm.

To complicate matters further, in formal feminist clinics labor was explicitly marked as either voluntary or waged. In such clinics, radical ambitions collided with pernicious classed and raced labor politics, which at many clinics could produced a hierarchy of white directors at the top whose dedicated labor was formulated through ethics of self-sacrifice and founding mothers, and more racially diverse staff working for low wages. The excitement of politicizing the microdetails of practice could easily transform into the drudgery of "shitwork."[121] Symptomatically, tensions over structure and hierarchy erupted on the pages of the feminist periodical *Off Our Backs* amid a heated dispute between the directors and paid staff at the Los Angeles and Orange County Feminist Women's Health Centers, leading to a staff "walk out" in 1974.[122] At least a few clinics attempted to resolve these tensions by appealing to "sensitivity training" from human relations research.[123] A retrospective sociological national survey of early clinics found that full-time staff typically worked sixty hours a week, for wages ranging from sixty to a hundred dollars—hardly a living wage.[124] Working in a clinic necessarily confronted antiauthoritarian protocols with the lived realities of devalued labor.

Likewise, antiauthoritarian protocols that governed the staffing of clinics by nonprofessional women who would be peers to clients were also extremely difficult to sustain. For clinic services to qualify for payment from Medicaid or insurance companies, they needed to meet state-licensing criteria that challenged their investment in nonexpert labor. From its inception, the Los Angeles clinic was under police surveillance, culminating in a raid broadcast live on television in which charges of practicing medicine without a license were laid for the act of applying yogurt to a woman's cervix.[125] The California Board of Medical Quality Assurance made numerous attempts to quash the practices of clinics, which by 1984 under the threat of losing access to state funding of family planning, led to the abandonment of lay health workers by Californian feminist clinics. In other states, similar clashes between feminist collective protocols and professional licensing occurred. For example, the Women's Community Health Center in Cambridge, Massachusetts, took over two years to get city and state approval. As a result, it ceased to operate as a collective. Reflecting on its struggle, the staff's annual report of 1978 remarked on the "ways bureaucratic red tape can be used for conscious harassment."[126] The clinic closed in 1981.

Thus, for clinics with feminist self help ideologies, the politicization of protocols over time became focused increasingly on the clinical moment, and not the organizational structure of labor. Struggles over funding and structure were exacerbated in the early 1980s by the rise of a militant antiabortion movement that not only obstructed the day-to-day work of clinics but also, at its most extreme, committed acts of violence including vandalism, arson, bombing, and even murder. Patients and staff alike had to walk through a gauntlet of antiabortion protestors before entering a clinic. Acts of arson, including bombing, tended to be underinvestigated, and legally were not treated as acts of terrorism. After an arsonist burned the first Feminist Women's Health Center in Los Angeles in 1985, the center never reopened. Within the Federation of Women's Health Centers, Redding and Chico were particularly hard hit, with daily picketing, regular vandalism, and phone threats. As beacons for trouble, both clinics were evicted from their original locations. Bombs, threats, arson, and even shootings became daily worries and regular occurrences.

As a result, the vision of a feminist health clinic as a microcosm of a politicized sociality and alternate technoscience waned. The clinic was no longer oriented as an equalitarian collectivity, but toward the security of staff and clients, as well as simply keeping its doors open at all. Under siege, the era of radical experimentation swiftly ended. The politicization of protocol feminism as a kind of expansive counter-conduct declined, often narrowed to the "quality" of service provision, and under the rubric of quality mainstreamed into family-planning services, such that protocol feminism would travel on transnational currents of foreign aid, no longer tethered to its radical, "revolutionary" roots.

Conclusion: Recursive Reassemblies

Feminist self help, as a protocol feminism, was not only fashioned out of strategic appropriations; its ethos of reassembly in each location allowed its protocols to move through a complex national (and transnational) itinerary of varied politicization. Local stratified histories—in Boston, Gainesville, Atlanta, Santa Cruz, and LA—were joined by road trips on interstate highway systems, telephone networks, mimeographed or photocopied pamphlets, manifestos, and periodicals transmitted through mail, cheaply made slide shows made possible through popular commercial photography, and mass-produced medical commodities that could be purchased in Oakland or in Kansas. Literal cut-and-paste rearrangements of pamphlets and texts—a standard practice—wove a dense set of citations across distant metropoles, assembling various forms of the imagined "we." While the West Coast Federation of Feminist Women's Health Centers was bound both by coastal highways and a shared "Black Book" of best practices, infrastructural conditions on local, national, and transnational scales required the repeated remaking of feminist self help.

The Black Women's National Health Project (BWNHP) is one example of the shifting biopolitics of feminist self help in the 1980s. Byllye Avery, a founding member of the Gainesville, Florida, feminist clinic, helped to start the BWNHP in 1981 in Atlanta as an explicit reassembly of feminist self help.[127] The BWNHP can be situated both as a continuation of the critical disidentifications and necropolitical analyses of many black feminists of the 1970s and early 1980s, as well as a narrowing of intervention through protocol and feminist self help to the ambit of health care. At its inauguration, the BWNHP developed a consciousness-raising program called "Black and Female: What Is the Reality?" Eschewing a focus on grassroots consciousness raising and gynecology, the first BWNHP meeting was instead organized as an antiracist "sensitivity-training" conference, while health issues such as mental health, heart disease, diabetes, and obesity that were prevalent in black communities were linked with the material conditions of a racist and sexist society, including personal and community experiences of violence, rape, and incest, and the internalization of oppressions.[128] From its inception, the National Black Women's Health Project called for black women to "love" themselves and "breathe life" into themselves.[129] Self help as expressed in the 1980s by the BWNHP had less a focus on changing clinical practice—nor critiques of political economy for that matter—but rather tended to concentrate on creating registers of valorization, positive emotion, and attending to health self-care generally (not reproduction specifically) in the face of necropolitical health effects.[130] In this way, the BWNHP both brought critical analyses of race to the protocols of feminist self help and constrained the scale of political action from sexed and raced political economies to self help protocols.

Through its explicit ethic of reassembly, feminist self help, though un-raced in its early LA articulation, could be reassembled as a raced politics in Atlanta, as an HIV clinic in Nigeria, as a gendered-citizenship project in Brazil, or as a project of continuous motivation in Bangladesh.[131] At the same time, the travels of protocol feminism expressed a particular version of politics that was both an exciting politicization of the details of technoscience, and a thinning of reproductive politics (or health) to the scope of the clinic and the body. Its flexibility, as later chapters will show, would allow practices of feminist self help to circulate across circuitous and contradictory transnational routes.

As a result of the violent attacks on feminist abortion clinics, and the "siege mentality" it produced in clinic workers, the 1980s saw the end of an era of radical experimentation through protocols. Yet it also meant that the forms of feminist self help practice that survived into the neoliberal 1980s were more open to appropriation during a decade in which the NGO became a dominant unit of feminist activity. Sonia Alvarez describes this process of NGO-ization as a feminist *boom*, a term which points to both the explosion of kinds and sites of protocol feminisms, but also implies a destruction wrought in that explosion.[132] As feminists of many stripes sought to appropriate funding and resources within the tentacles of the family-planning, public health, and development industries, the participatory core of self help protocols could be rephrased as an injunction to responsibilize oneself without politicization. In neoliberal formations of self help, participation, individual responsibility over health, and information exchange were no longer just forms of counter-conduct; they were conventional requirements for access to health care infrastructures.

This story of feminist self help, as an example of the entangled politicization of technique and life, is not just about the history of feminism. It is also about the history of technoscience. While critiques of science within feminist self help tended to deploy the straw man of an unemotional and authoritarian science, the protocols of feminist self help—such as its valuing of emotional labor, politicization of research, and attention to the details of practice—share their features with other kinds of late twentieth-century science, features that science and technology studies scholars have pointed to in our current historical conjuncture.

Science and technology studies as a discipline emerged in the moment of feminist self help. It too heralded the presence of politics within the daily nitty-gritty of technoscientific practice. While self-reflexive critical projects are instances of critical diagnosis, they are also perhaps just as much symptoms of their moment, symptoms of a larger late twentieth-century legibility of politics as constitutive of technoscience, a legibility that exceeded the bounds of rarified academia.

CHAPTER 2

Immodest Witnessing, Affective Economies, and Objectivity

Sit in a circle. Assemble a kit composed of mirror, light, and plastic speculum for each participant. Lubricate your speculum with the duckbill closed
and the handle in the upward position. Insert with care. Squeeze the
handle and press down. You will hear a click to let you know it is locked
open. To see yourself, hold the mirror between your legs and direct the
light toward it. The light will reflect off the mirror into your vagina so that
your cervix will pop into view. Enjoy the lush color, texture, odor, and
shape of the cervix and vaginal walls. Take turns sharing your observations with the group. Admire the subtle variations and the fine differences
in form. Track changes.

These are some of the ingredients making up the feminist protocol of
vaginal self-exam in Los Angeles in the 1970s. This assemblage of commercially available devices, behavioral scripts, affective economies, and embodied subjects became feminist self help's iconic practice. Beyond a health
care protocol, I want to argue, vaginal self-exam exemplified a historically
particular and politically charged refashioning of *objectivity*. At stake in
the protocol of vaginal self-exam was how to see and to create knowledge
about health and bodies. In other words, vaginal self-exam made manifest the epistemological stakes—the politics of how-to-know—crafted
into the biopolitical project of feminist self help. Moreover, feminist self
help practitioners argued that embodied ways of knowing produced better
knowledge. While historians of science tend to look to professionalized

disciplines—physics, biology, astronomy, statistics—or increasingly to high-tech domains—such as genetics and nanotechnology—to historicize modes of objectivity, this chapter takes up the practice of vaginal self-exam to draw attention to how the history of objectivity in the late twentieth century was also crafted through politicized interruptions by nonprofessionals and, more specifically, by lay researchers who situated themselves as the embodied "subjects" and "objects" of technoscience.

While it might be tempting to dismiss vaginal self-exam as simply an expression of a privileged narcissism masquerading as radicalism in the 1970s—titillating but ultimately trivial—I want to take this practice seriously as a historically specific and influential form of technoscience that reassembled objectivity, thereby examining it in the same way I would any other manifestation of technoscience. Moreover, this chapter takes seriously vaginal self-exam as a juncture both in the genealogy of objectivity in science *and* in the genealogy of the contemporary field of science and technology studies (of which this book is a part) that has examined "objectivity" as historically particular rather than transcendent. Science and technology studies scholars have long been concerned, even obsessed, with questions of objectivity.

The historians Lorraine Daston and Peter Galison, in particular, have tracked "objectivity" as a changing epistemic virtue only crystallized in the nineteenth century as the valuation of "blind sight" in observation— as valuing methods for capturing the world in ways that minimized or circumvented the imposition of "subjectivity."[1] Daston and Galison, in their account, pit the epistemic virtue of blind sight objectivity against eighteenth-century practices of "truth-to-nature" that sought to capture the ideation, the perfect version, of the object under study. For example, an eighteenth-century natural historian was not just interested in illustrating any flower as an example of its kind, but instead tried to discern an abstracted perfect form of that kind of flower only after studying many examples. Yet other genres of objectivity, in contrast, aspired to let "nature speak for itself," requiring the scientist to preserve the imperfection and variation in the object by virtue of training his or her will against imposing interpretation on data. How to get nature to speak? Through what practices and what modes of representations?

The history of objectivity does not offer up a narrative of the progressive arrival at a best objective mode. Instead, scholars have charted a shifting proliferation of practices and values through which scientists have aspired to objectivity, such as mechanical objectivity, in which the use of an instrument aspires to bypass subjectivity; or structural objectivity, in which scientists sought to observe that which was invariant and, thus, structural to the world.[2] Efforts to historicize objectivity also involve attending to the history of its other—subjectivity. The kinds of selves and "subjectivations" available to scientists profoundly shaped how the self was seen as either a hindrance or a help to knowledge production. In this dance between objectivity and subjectivity, Daston calls our attention to the "moral economies" of science—historically specific "dispositifs" of affect-saturated values that shape what counts as good science.[3]

In the twentieth century, multiple modes of objectivity have not just replaced each other; they have accumulated. In medicine, for example, X-rays, stained slides of tissue samples, and the graphical representations of EKGs—each in turn calling on different versions of objectivity—all required trained expert interpretation. The mechanical objectivity of instruments became the infrastructure on which trained diagnosis or interpretation was then made. The doctor, as well as the scientist, in the late twentieth century was not necessarily self-removing, and instead could be heralded as holding a trained form of cognition. Mental abilities to identify patterns and notice anomalies or develop theoretical explanation were celebrated as an epistemic value—a kind of highly valued labor.[4]

By the time feminists in Los Angeles began practicing vaginal self-exam as a new clinical and research protocol, science and medicine were in practice removed from the iconic figure of the lone brilliant scientist struggling diligently to disentangle the real from the representational, and the representation from his own bias. They were instead associated with large research programs at modern universities, hospitals, and corporations, with "big science" and "medical monopolies" that had complex labor stratigraphies. In California during the 1960s, protests against the ways university-based research was complicit in the Vietnam War helped to elicit passionate critiques of "neutral" science from many quarters.[5] Critics from the Left argued that good science should have applied benefit and be socially accountable; from the Right, they demanded that overindulgent technoscience be made fiscally accountable.[6] At this juncture, science and medicine were reinvigorated with other sets of epistemic values—particularly the capacity to be applied and made commercially viable. Nowhere was this truer than in California, where from computing to biotechnology, notions of social change were hinged to the "entrepreneurial energy" of a new generation of researchers. The subject-figure of the scientist could be celebrated as passionate, motivated, socially networked, energetic, and creative—virtues of entrepreneurialism.[7]

So too, in the 1970s, was the authority of physicians challenged by feminists and other patient groups, as well as by the transformation of medicine into "biomedicine," itself composed of large technical and corporate systems.[8] Movements to govern medicine through consumer choice further coupled with the rise of the pharmaceutical fix as the dominant form of late twentieth-century therapeutics. Laboratory tests offered numerical measures of unseeable, sometimes unfeelable, micrological phenomenon as a normative kind of symptom for diagnosis—from diabetes to cholesterol—providing quantitative indicators of the necessity of pharmaceutical intervention.[9] Medicine was becoming "biomedicine" as research and care collapsed into one another. Thus in the 1970s, circling through the cognitive labor of the trained expert, the entrepreneurial passion of the researcher, the personalized choice of the patient, and rationalized medical circuits of pharmaceutical research, the newly emergent moral economy of biomedicine assembled together a multitude of ways of knowing with a variety of subject-figures. Biomedical research spiraled through diagnostic protocols, into clinical judgment, and then through pharmaceutical therapy, which in turn called for yet more clinical data collection. These elements constituted a regime that sociologists Albert Cambrosio and Peter Keating have called "regulatory objectivity," in which trained expertise was just one component in a multisited, multimoded itinerary of knowledge making that brought together sometimes contradictory moral economies.[10]

This spiral of multiple modes of objectivity in biomedicine was a feature of a larger late twentieth-century change in which the epistemic virtue of letting "nature speak" came into conflict with the ethic of making useful, applicable, and instrumental. Instrumentality—what could be done—increasingly trumped abstract truth claims as a virtue. Science was remade as a "technoscience" that explicitly intervened in the world— treated it, engineered it, commodified it, built it, experimented with it.[11] Apprehension itself was not mere observation, it was world making as the very scientific acts of study involved altering the world. The objects of technoscience—in physics, biology, ecology, engineering, computing, and so on—were valued as the conjoined moment of apprehension and intervention.[12] For example, in the field of reproductive sciences, which in turn was a crucial source of "biotechnology," an experimental epistemic ethic sought not just to unravel the mechanics of reproduction, or disassemble it into micrological substrates, but to alter, modify, and render life technical, commodifiable, and governable.[13] It was during the emergence of this manifold landscape of often conflicting epistemic values that vaginal self-exam was crafted as a form of counter-conduct, a way of knowing explicitly created as a reaction to dominant practices.

In placing vaginal self-exam in this tangled genealogy of objectivity, I want to repose Daston's formulation of "moral economies of science" to suit the recent past of the late twentieth-century technoscience, including forms of technoscientific counter-conduct. Daston develops the concept with a sense of the eighteenth-century use of the word *economy*—as a balanced system of organized attributes. In contrast, I will rework the term in its present-day sense by attending to how affect and embodiment are, as well as being epistemic concerns, entangled with cultures of late twentieth-century capitalism. Hence, the notion of moral economies of science applied to eighteenth-century practices can be reworked as *affective economies of technoscience*, which refer to how capacities to feel, to sense, and to be embodied are valued within political economies of technoscience.

As an attempt to practice research as a political project that could tell better truths, feminist self help in the 1970s drew together an affective economy of technoscience that hoped to challenge dominant practices. Embracing instrumentality, feminist self help did not just seek to simply reveal a new truth about reproductive health, but offered new practices for interacting with, caring about, and managing reproduction—to seize the means of reproduction. Vaginal self-exam, as a protocol, explicitly attempted to operate outside of professional and profit-driven biomedicine, and hence grappled with the role of capitalism and authority in knowledge making by virtue of crafting alternative affective, embodied, and political, rather than economically productive, epistemic values.

The anthropologist of science Natasha Myers has developed the term *affective entanglements* to describe the reiterative "body-knowing," rich in affect and a sensory perspective, that contemporary molecular modelers have with their objects.[14] In Myers's analysis, these affective entanglements materialize molecules as much as make them knowable. In this chapter, I want to build on this sense of affective entanglements to go beyond describing the evocative attachments between researchers and the objects they engage, in order to capture the uneven *political economy* of affect. Vaginal self-exam, as generated in an affective economy of technoscience, worked to materialize bodies and researchers in new ways, on the one hand, and was caught between feminist counter-conduct and broader historical developments, on the other. As part of the history of objectivity, vaginal self-exam signaled the emergence of affectively charged practices as a core epistemic value. This value was both heralded by feminists committed to emotion and embodiment in knowledge making, and given monetized value in gendered labor and entrepreneurial technoscience at the end of the century.

Immodest Witnessing

The facilitator was the first person in a self help group to perform a vaginal self-examination, breaking the taboo of nakedness by nonchalantly removing her clothes and inserting a speculum, all the while providing a narrative of what she was feeling and seeing: "Here's my cervix, that looks like it usually does . . . the Os tips down, which is normal for me . . . I see some white secretions that don't bother me and tend to appear during the middle of my cycle . . ." and so on.[15] The facilitator's role was not simply to demonstrate the mechanics of opening a speculum or to point out anatomical parts; it was, more importantly, to model a procedure of observation for others to follow. The facilitator did not lecture or set out rules; instead she set the tone for the group by her example. The language of the facilitator provided a palette to see, feel, and speak with, and to bond around. Self-exam ideally sparked a mutual thrill, an "exhilaration" at the daring of taking technoscience into one's own hand, an affective atmosphere that was meant to enhance a feeling of solidarity. "Clicking of speculums, the buzzing of several conversations, and intermittent choruses of laughter" were the sounds of affective entanglements.[16] The guiding epistemic values within feminist self help protocols are captured here: using your body to know your body, valuing and producing affirmative affective relations, appreciating variability, and collective research.

The protocol for vaginal self-exam was disseminated in an abundance of instructional images in slide shows, mimeographed handouts, films, pamphlets, and books. The Los Angeles Feminist Women's Health Center, in particular, took hundreds of photographs, some of cervixes, some of genitalia, others of the act of vaginal self-exam and other forms of appropriated biomedical labor. Not simply a straightforward set of written directions, these visual practices and materials offered a particular way of visually manifesting a protocol as a practice of seeing. The images had tactical importance by "taking back" representations of female anatomy in a historical moment when new technologies of visualization extended the biomedical gaze and old anatomical illustrations erased sexuality and particularity.[17] Moreover, I want to argue, such visual practices were constitutive of a reassembled status of the subject in objectivity. I will call this new subject-figure the *immodest witness*.

In tracking the figure of the immodest witness of vaginal self-exam in the visual productions of feminist self help, I am inspired by the "material-semiotic" figures in the scholarship of Donna Haraway, where she takes the Cyborg, the OncoMouse, or the Modest Witness as oppositional, and yet noninnocent, means to query technoscience. For Haraway, these are "performed images that can be inhabited."[18] What I am calling the *immodest witness* is likewise a complex oppositional and yet entangled subject-figure, incited into being not only in images, but also in practices, bodies, and affects. The visual tropes of vaginal self-exam functioned as procedural instructions, yet also as a generative reassembly of subjectivity and objectivity through embodiment. Starting with yourself in what you were studying, and highlighting your affective entanglements, were epistemic values that aspired to produce better, more accurate, knowledge. In addition, this vantage point promoted entanglements that offered a version of the scientist-subject as deeply responsible and implicated in her object of study. Simply put, who you were affected what you could know. For the immodest witness, subjectivity was not an abstract problem of seeing but a question of concrete and particular embodiment that promised a better—a more proximate and intimate—route to objectivity.

Since the seventeenth century, the subject-figure of the "modest witness"—who aspires to hold the personal details of their subjectivity, status, class, race, gender, religion, and mood apart from observation—has been a recurring figure in scientific moral economies of blind sight, or what Haraway called "the view from nowhere" within what Sharon Traweek called "the culture of no culture."[19] In contrast, a naked woman observing herself was an *immodest* witness who was not only embodied through her eye, but materially displayed her embodiment as a constituent component of observation.[20] The contrasting figure of the modest witness had its origins in the experimental sciences of the seventeenth century.[21] Gendered male, raced European, and enjoying the status of gentleman, the subject-figure of the modest witness could be trusted to make reasonable observations, adding nothing beyond honor from the specificity of his own person.[22] The epistemic virtue of modesty involved making oneself humble, using one's senses as simple instruments, and fashioning oneself as the ventriloquist for the objects studied. At the same time, experimental practices could encompass using the "self-evidence" of one's own body as a kind of sensitive medium, strained of the particular.[23] This immodest witness, then, can be situated genealogically as a twentieth-century re-iteration of various figures in the history of science who sought in their own sensory, cognitive, and emotional experience the empirical basis of their research, suggesting that recourse to the body is a recurring theme in scientific praxis when other modes of apprehension are deemed inadequate. Within this lineage, however, not just any person could offer the self-evidence of discerning, trustworthy sense. The "modern" European modest witness was a subject-making figure that crucially delineated the kinds of persons who could (purportedly unmarked subjects) and could not (marked subjects) credibly produce knowledge.

Historians of early modern science have excavated a host of other kinds of labor involved in producing experimental and observational sciences—labor less valued in scientific production and typically rendered as technical, rather than intellectual, work—from instrument makers, to family observatories, to illustrators, to local inhabitants who assisted collectors, to fieldworkers. Thus, the subject-figure of the modest witness, with abilities for reasoned judgments, was supported by larger colonial, gendered, raced, and classed figurations of what kinds of bodies were seen as having capacities to reason, and what kinds of labor were truth producing rather than merely instrumental.

Immodest witnessing, in contrast, was explicitly both an object-making and subject-making process that elevated the layperson as expert in the particularities of herself. I use the word *immodest* here to draw attention to the project of laying bare the importance of the subject in knowledge making, and of challenging notions of chastity and modesty that prevented women from displaying, valuing, or studying the female reproductive body, or even marking the subject-figure of the scientist as sexed, and hence as a particular, not abstract, person. The visual practices of vaginal self-exam boldly announced the sexed embodiment of the laborer in knowledge production. For practitioners, the immodest witness was part of a tactic of "demystification" concerned with unmasking the craft of knowledge hidden by professionalism, thereby drawing attention to who was allowed to participate in the labor of science, revealing what had previously been obscured as actually the product of relations of power. In its task of demystification, feminist epistemology was indebted to Marx's call to demystify the commodity as actually composed of social relations and labor.

At the same time, in their elevation of experience and sensation as epistemic virtues, radical feminists tended to consider their knowledge-making practices as a return to an empiricism associated with the scientific revolution: "The decision to emphasize our own feelings and experiences as women and to test all generalizations and reading we did by our own experience was actually the scientific method of research. We were in effect repeating the 17th century challenge of science to scholasticism: study nature, not books, and put all theories to the test of living practice and action."[24] In this way, the feelings and experiences of immodest witnessing became a primary passage point through which the validity of already existent knowledge—such as Marxist theory or biomedical descriptions—had to be tested.

The immodest witness was cleverly captured in the canonical self help image of a woman examining herself with a mirror and a speculum (see figures 2.1 and 2.2). Unlike contemporaneous drawings of pelvic exams in gynecological textbooks—typically either a straight view into the vaginal canal, evoking the camera angles of pornography, or a cross section of disembodied organs with arms, legs, and head severed—images of the immodest witness put the viewer in the eyes of the woman examining herself. Our gaze is taken over our own pubis and into the mirror we are holding between our legs. In the mirror, the speculum guides our gaze to the cervix, yet the mirror as symbol of a transparent access to the world is resisted, for the illustration makes us aware of the mirror's frame and interpellates us into our own embodied gaze. The sex of the observer could not be missed.

Acts of women studying their sexed bodies through their bodies created a recursive circuit that joined the observer and the observed in a single gesture. This conjoining, first, rendered the body under observation an object of inquiry active in its own observation and, second, rendered the observer an embodied figure entangled with the object under study. Hands were an important trope in the figuration of the "möbius"

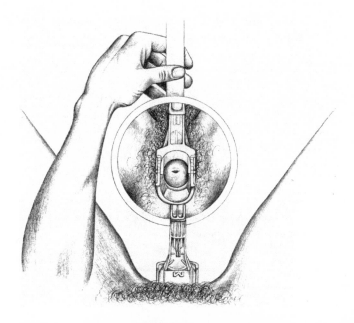

2.1. Illustration by Suzanne Gage of vaginal self-exam, organized as an act of immodest witnessing. Following feminist self help methodology, the illustration situates the viewer as if looking into their own body. From FFWHC, A New View of A Woman's Body (1991), 24. **2.2.** Slide of immodest witnessing taken by Lorraine Rothman for the 1971 feminist self help road tour presentation. The production of the photograph, in which the photographer images herself, reinstantiated the self help method being represented. Courtesy of Lorraine Rothman.

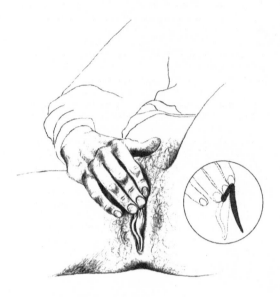

2.3. Active hands and probing fingers were an important trope of feminist self help imagery and the invocation of the immodest witness. Here a woman is using her hands to roll the shaft of the clitoris in an illustration by Suzanne Gage to represent the findings of the 1978 Clitoral Study. From FFWHC, *A New View of A Woman's Body* (1991), 36.

agency of the immodest witness, intended to convey the use of the observer's senses, the generative lushness of the body itself, and the agency of the woman being examined. For example, hands figured prominently in the feminist self help project of crafting a "new definition" of the clitoris, which dramatically expanded the anatomical scope of the clitoris, as well as provided a sense of its detailed function. Illustrations of clitoral anatomy (drawn by Suzanne Gage, who developed much of the visual vocabulary for feminist self help) began with four drawings of fingers spreading the outer lips of a vagina, pulling back its hood, rolling the shaft, and squeezing the glans (see figure 2.3). In contrast to the classic European sculpted figures of Nature opening her clothes to reveal her secrets to the gaze of an unrepresented scientist, the immodest witness was actively manipulating and probing her own body, positioning herself as both affectively engaged and the active creator of knowledge. In general, if feminist images did not portray woman's hands as actively probing cavities or pulling back folds, they were still often visible on her legs or stomach, signaling that the body under examination remained an active part of the procedure.

The feminist self help work on the clitoral study of 1978 was some of the most sexually charged research conducted, in which the process of orgasm itself was studied in a group.[25] The results of this pleasurable labor offered a dramatically enlarged clitoral anatomy, as well as a detailed description of its function filled with intimate affective observations at the crossroads of hands and genitals. The revised clitoris was brought into lively and timely embodiment through pleasure, touch, and sight in both the registers of practice and representation. This affect-drenched research into genitals, enfolding the feeling object with the desiring subject who disidentified with norms, was distinctly a form of queered research. While introductions to vaginal exam were able to avoid questions of sexuality as the signature upturned speculum handle obscured the clitoris from the scope of introductory observation, "advanced groups" were much more likely to sexualize their affective economies of research. Advanced collectivities, for example, took up questions of ejaculation and lesbian health in intensely affective circumstances.[26] Thus, the immodest witness was also potentially an explicitly queered subject, who violated heteronormativity, not only by assuming the status of the scientist, but also by virtue of the affectively charged same-sex circuits of sensory observation of parts of the body deeply saturated with sexuality.

Affective Economies and the Not Uncommon

While vaginal self-exam invested this circuit of self-study with an epistemic privilege, the immodest witness was also produced in an assemblage made possible by, and drawn into new forms through, larger affective economies. The politics of immodest witnessing is inseparable from other late twentieth-century affective economies which incited feelings, sensations, and embodiments as valued and devalued aspects of more dominant research cultures. Integral to Daston's notion of moral economies of science is an insistence on studying morals, not as residing in individuals, but as historical formations that individuals come to inhabit.[27] Relatedly, the feminist theorist Sara Ahmed argues that we should think of emotions as constituted in "affective economies," in which emotion does not reside as a primal force in the individual or the psyche; rather, she presents it as incited through historically specific arrangements of circulation between subjects.[28] Unlike Daston, for Ahmed, the notion of "affective economies" has a distinct affiliation with how commodities are analyzed by Marx. Just as Marx shows that the commodity, which seems a thing in itself, conceals the social relations and labor that make it possible, for Ahmed affects as emotions, which seem to originate prior to social relations in bodies and minds, are better understood as generated in historically specific circulations that align and differentiate bodies in particular ways. Just as the commodity's value is the effect of circulations of labor, money, and capital, the value of certain affects—happiness, passion, or even hate—likewise are the accumulated effects of patterns of circulation between subjects in uneven worlds and thus perform particular kinds of work arranging social relations. In other words, value is produced, assigned, and circulated in affective economies.

Feminist self help did not use the term *affect*. It did, however, use the term *experience* as an emotional, sensual, and embodied value. Therefore, it is important to think through the question of affective economies in immodest witnessing by first considering the epistemic privilege with which feminist self help, as with radical feminism more generally, imbued "experience." Perhaps the most crucial axiom of consciousness raising and feminist research was that all knowledge production should begin with women's experiences. With feminist self help, however, experience was both the empirical material analyzed (the embodied experience of being a "woman") and the immediate encounter with one's body produced through vaginal self-examination (the experience of looking at oneself). At work in statements such as "I saw this," "I was there," "I felt that" uttered at feminist self help meetings was the assertion of a purported epistemic privilege gained from the immediacy of observing one's self. It shouldn't be surprising, then, that the movement's literature is loaded with the term *experience* and that its uses were both tangled and polyvalent.

The experience of viewing one's body was heralded over and over as an affect-laden and consciousness-raising event. For example, the "Self Help in a Suitcase" road show, which ended with a demonstration of vaginal self-exam, prepped its audience with a slide show photograph of an immodest witness staring over her trunk, through her legs, and into the smiling faces of a multiracial group of women looking back (see figure 2.4). The script for this slide declares, "Happiness is knowing your own

2.4. An image from the slide show used in the traveling feminist self help "road show," exemplifying the affective economy of vaginal self-exam. The accompanying script for this image was "Happiness is knowing your sister's cervix." Courtesy of Lorraine Rothman.

cervix. . . . Happiness is knowing your sisters' cervix."[29] In moments like these, feminist self help fostered an affective economy that sought to align women in a collective project of bonding and pleasure through the epistemic value of collectivized observation. In the 1970s, the act of vaginal self-exam was believed to have a powerful consciousness-raising effect that is often hard to imagine today. Downer described its effect as "that shock, that extremely rapid rise of consciousness." "The women who did self-exam, one second they were not able to open their legs and the next there it was, and no big deal. I mean it was that fast. It was extremely exciting."[30] The act of vaginal self-exam was understood as prompting intense emotions as well as aligning and bonding participants. Crucial to this effect was the small group format.[31] Self-examination, despite its name, was not an exercise in individual self-reflection. Alone, it was easy for someone to perceive her genitalia as strange or her reproductive and sexual life as pathological. Through comparative analysis in a group, women were called upon to translate their individual experience of looking at their cervix into information about "women" as a class. Group acts called on women as part of a politics and a "movement"; single acts stayed "within the confines of her own four walls."[32] In other words, consciousness raising can be seen as not just assembling a group discussion, but as constituting an affective economy that aligned participants in relationship to each other, hailing women already bonded who were "sisters," as subjects nominated into affective relation by virtue of mutually prompting the recognition and "validation" of each other's observations. Hence, the möbius joining of subject and object in acts of vaginal self-exam was further animated in the group by aligning women with one another through a shared emotive practice of mutual validation.

"I think it is a very sound scientific principle," explained Downer. "We validated every woman's experience. That was our means of learning. Whatever everybody said was what it was. Not what we had read about."[33] It is important to note that the purpose of the self help group was not to collectively assent to a common truth claim, because each woman was expected to have her own individualized sensations, her own "experience," and only she was authorized to make judgments about her own body. Within collective immodest witnessing, participants "validated" their observations by affirming, rather than the deindividualized objectivity of a fact, each woman's capacity to judge herself, within her own skin. Thus, the other women participating in an instance of vaginal self-exam were also immodest, in the sense of being entangled with each other rather than impartial to their objects of concern. Consciousness raising, in this sense, was not just an analytical, political, and social technology, but also an affective technology. It set protocols for using language and for interacting, as well as for feeling, aligning, and bonding.

Though it may be tempting to take "experience" as a self-evident originary point of explanation—as that which explains, not that which needs to be explained—I want to attend to how claims to represent experience operate by taking as given and already constituted the identities of those whose experiences are being represented; whereas the task of the critical historian is to excavate the production of subjectivities though the ways the evidence of experience is imbued with an authentic primacy.[34] In other words, the "evidence of experience" needs to be historicized. Feminist self help nominated the evidence of experience as an affectively entangling truth claim that worked to actively incite, align, and differentiate subjects and not just represent them. In feminist self help, then, the evidence of experience worked within a particular economy of knowledge, and in so doing both heralded affectively entangled subjects, and posited affective entanglements with the ability to create a better mode of objectivity.

The question that I am interested in here is not whether experience rightly has epistemic privilege—has a greater access to better truths; rather, I want to ask, What were the practices by which "experience," as a kind of affect-saturated and embodied mode of objectivity, was materialized? And further, what ranges of phenomena were rendered perceptible and intelligible by collective economies of intimate attention to embodiment? What range of political tactics were performed when the qualities and scales associated with embodied and affective experience were inscribed into the methods for apprehending the world?

Typically, women took turns describing their experiences on a particular thematic. These collected experiences were then analyzed, not to provide therapy for the individual (though it often had a purported therapeutic effect) but to chart the social conditions common to women as an oppressed group. The very act of speaking experience could be arduous. Meetings could be long winded and personal disclosure agonizing, participants sometimes had to find the words or invent the neologisms to express what some called "the problem that has no name," and, most significant, affective entanglements did not necessarily work so smoothly and joyfully.[35] Consciousness-raising groups could result in bitter feelings or feuds and create relations of alienation or disidentification, not only circuits of bonding and collectivity. Because consciousness raising was a technical process, the critical potential of "experience" as a kind of evidence was not assumed to be self-evident; instead, it was understood to be won through hard group work. For example, that one experienced patriarchy was not necessarily evident to the lone woman, yet through the laborious efforts of consciousness raising, personal experience was to be transformed into the evidence that could aspire to an analysis of structural oppression. In practice, this earnedness was not always foregrounded, and could even be overlooked as a group slipped into navel gazing.

The now worn phrase "the personal is political," coined by radical feminists in the late 1960s, was meant to signal the politicization of that which was previously held as personal, individual, and even trivial, not the personalization of politics into a private domain of self-improvement.[36] And like consciousness raising, the first lesson one learned in a self help meeting was that of commonality: "What you thought was peculiar to you was in fact shared by everyone."[37] At the same time, the slogan "the personal is political" captured a danger within the method: the insight that social structures manifested themselves in what seemed like idiosyncratic personal events could be used to elevate quite historically and geographically particular insights into problematic universals.[38]

Consciousness-raising manifestos of the late 1960s and early 1970s—a distinctive genre of the time—had varying notions of their epistemological achievements.[39] For many white feminists, consciousness raising was thought to reveal "what you thought was peculiar to you was in fact shared by everyone."[40] In Pamella Allen's well-known methodological handbook *Free Space* (based on the efforts of the San Francisco group Sudsofloppen and also her work in New York Radical Women), she provided an affective recipe for the successful consciousness-raising group: "Not only do we respond with recognition to someone's account, but we add from our own histories as well, building a collage of similar experiences from all women present. The intention here is to arrive at an understanding of the social condition of women by pooling descriptions of the forms oppression has taken in each individual's life."[41] By "pooling" and "collaging" experiences, many radical feminists strove to find an undergirding system of oppression that could explain women's varied experiences; though the exact form this oppression took in each woman's life varied, these differences were assumed to fit together like pieces of a puzzle, revealing the workings of a patriarchy beneath.[42] In this base-superstructure model, for example, patriarchy was the common structural ground and women's lives the varied expression. At the same time, this variety was nonetheless bound together through the common category of "women." In the 1971 preface to *Our Bodies, Ourselves*, for example, the collective stated, "In some ways, learning about our womanhood from the inside out has allowed us to cross over the socially created barriers of race, color, income and class, and to feel a sense of identity with all women in the experience of being female."[43]

Though this sense of "the shared" as consciousness raising's product was perhaps most typical, other feminists invested consciousness raising with different effects. For example, the Combahee River Collective of Boston (discussed in the previous chapter), began as a consciousness-raising group that sought to expose the multiple and contradictory "interlocking oppressions" that not only aligned subjects, but divided subjects against each other and within themselves. Thus, consciousness raising—as a method for producing and analyzing the "evidence of experience"—could also be mobilized to analyze the production of difference and contradictions, and it was only some versions that highlighted commonality.

Vaginal self-exam, and feminist self help more generally, were yet another iteration and rearrangement of consciousness raising. The role of the evidence of experience in vaginal self-exam differed from conventional consciousness raising in that it included the "immediate concrete" moment of examining one's own body.[44] The affects involved were not generated by reflecting backward in life stories or inward to psychic states. In contrast, they tended to come from the immediate sensations of the exam itself. The evidence of experience joined together past reproductive, sexual, and medical events with the immediate moment of self-exam, which involved both the sensations of the speculum and the affective relations involved in the group, as well as the sensations of observation—what one saw, smelt, tasted, or felt. Immediacy was conveyed through rich sensory narratives: the feeling of pressure as a speculum clicked into place, the pinkish color of the cervix with or without reddish hues, the moisture or dryness of the vaginal canal, the sweet or musky smell of secretions, the look of the curly or toothy flesh of a hymen. A woman might even taste the sticky residue left on the speculum once it was removed. The fine-grain description and even effusive language fostered a distinctive aesthetic sensibility that marked each woman's cervix and vagina as unique, often likened by self helpers to the individuality of human faces. The sharpness and texture of the observations materialized a lush array of small incidental individual differences. Biological variation—idiosyncratic health histories and anatomical quirks—were the incidental experiences to be gathered through a fine-grain corporeal attention. When anatomical variations were collected in the self help clinic, feminist self helpers pointed to a shared reproductive body underneath—"below the waist and above the knees"—but this shared domain was nonetheless lively with variation.[45] Their intimate examination of reproductive variation was not primarily a search for ill health; in contrast, it was an effort to remove reproduction from its association with pathology—"taking the routine into our own hands"—and revaluing embodiments in terms of, not in spite of, individual biological deviations.

Further, instead of a straightforward search for the undergirding common that characterized much consciousness raising, vaginal self-exam was crafted as a means to recognize that the "irregular is not uncommon."

This double negative of "not uncommon" is crucially different from "common."[46] Vaginal self-exam, as well as the research in "advanced groups," sought to study the not uncommon as routine anatomical variation, as ordinary secretions, and as ubiquitous infections so that these phenomena could be depathologized and seen as more appropriate to "home care" by women themselves than to medical care by doctors.

Attending to the not uncommon of reproductive health was likened to oral health practices—in terms of teeth brushing, gargling, self-inspection of the mouth. Both were practices done outside of medicine. Authority to judge one's own vagina and thus "demystify" reproductive anatomy was made analogous to the unexceptional act of examining one's own mouth. Unlike an organ—a technical term that might spring to mind when hearing the gynecological term "internal exam"—the mouth was an accessible cavity laypeople regularly inspected, took care of, and treated.[47] Both were "open to the outside" with mucous membrane linings; neither were a sterile environment.[48] Many things are put in the mouth, and so too with the vagina: fingers, penises, tampons, spermicidal foams and jellies, diaphragms, douches. And other things came out, not least of which were babies. Thus, according to self help protocol, a woman should feel licensed to have the same access and relationship to her vagina as she does with her mouth. "It seems odd, indeed," instructs the manual *How To Stay out of the Gynecologist's Office*, "that the same woman who would not dream of going to a physician for a sore throat spends time and money in visits to the physician for vaginal infections that she could treat herself."[49] Home-treating one's vagina was likened to taking a throat lozenge.[50]

At the same time, the "not uncommon" was also a valuation of variation itself. Variation was its own epistemic virtue and, moreover, variation gave the evidence of experience a particular form, one which was concerned with searching for and positively appreciating idiosyncrasies. To this end, over the course of the decade the Federation of Feminist Women's Health Centers (FFWHC) took hundreds of pictures of genitals and cervixes, recording a lush field of individual variety. In this way, so called not uncommon problems were refused the label of pathology or deviance, and instead were heralded as unexceptional variations that nonprofessionals could recognize, monitor, and manage.

The attention to the primacy of emotions and the "immediate concrete" of the body can be understood as more than assigning epistemic values; it materialized the very phenomena that it imbued as originary. In other words, affects and bodies were not just studied; they were materialized. I use *materialization* here to name the ontological politics through which phenomena are given specific form as material, "real" entities or relations with particular boundaries, qualities, and durations and, moreover, as phenomena understood as outside of history, as primal, or prior to historical constitution.[51] Emotions, sensations, and flesh are all drawn through this kind of ontological status, as phenomena granted both a material status and a primal realness. As the feminist theorist Judith Butler argues in terms of the materialization of what counts as matter, such designations of primal realness are themselves "the effect of power, power's most productive effect."[52] In other words, it is precisely at those conjunctures in which affects and bodies are imbued with ontological priority—by science, but also by feminists—that can be read as the political effects that require historical excavation.

By attending to the personalized and interindividual difference within a group as the not uncommon, vaginal self-exam offered a mode of collecting data that called into being what I will call an *ontological collectivity*—a materialization of a continuous field of difference rather than a fixed form. That is, interindividual differences were not judged by an abstract norm, or in terms of a fixed fact; rather, they were assembled into a collectivity of living variation both between bodies and within any given body. Thus, for feminist self help, the category "woman" was not absolutely unitary; instead it was a living ontological collectivity across bodies and over time. While feminist self help aspired to hail each woman as equally knowing subjects, at the same time it insisted that each woman would know herself—her difference—differently: only she could have the epistemic privilege of knowing her personal and unique affects and sensations that in turn would be added to the ontological collectivity.

The visual vocabulary of feminist self help images captured this epistemic value of variation. Most images were not only of embodied women; they were of the body of a particular woman, who might sport a pair of glasses, have scraggly pubic hair, or slouch[53] (see figure 2.5). The women represented in these images were clearly raced, diverse, and individual. Specificity mattered, and immodest witnessing sought to carefully attend to the specificities of the individual body as well as the variations between bodies, corralling these variegated bodily expressions into the ambit of the figure of the "well-woman."[54] Thus the immodest witness was the exemplar of a quite remarkable reassembly of objectivity that altered,

2.5. Suzanne Gage's illustrations tended to portray particular, ordinary, and diverse women, not abstract women. From FFWHC, *A New View of A Woman's Body* (1991).

one, the status of the subject (as particular, embodied, and affectively entangled with its object of study) and, two, the epistemic values of observation (valuing variation, sensation, and emotion individually and collectively).

At the same time, attention to the not uncommon asked participants to recognize themselves in each other according to the protocols of consciousness-raising group work. In fashioning this web of mutual summoning, practitioners were not simply discerning common patterns; they were evoking each other as politicized and connected subjects. Speaking across a circle, women recognized each other as embodied agents capable of truth claims. "Responding with recognition" thus asked women to valorize one another as highly individualized truth-tellers and to align as part of a politically charged cohort of "women," a more abstract commonality. Thus, the importance of individuality to the ontological collectivity created by vaginal self-exam was in tension with the necessity of invoking woman as a universalizable sex.

Supporting this tension was a tendency of feminist self help groups to be composed of similarly located women. While self help groups could gather around more specific alignments (as lesbians, as menopausal women, as black women, and so on), a tendency to create affective entanglements across similar locations was supported by attention to the not uncommon. For example, the highly individualized immediate evidence of immodest witnessing was hitched with the politically charged class of "woman," resulting in a purposeful erasure of the more complex and contradictory identities that cut across woman as a kind. Moreover, the attention to unraced and individualized biological variability could support the suspension of questions of race in the name of a flexible antiracism and, hence, implicitly reinforced the unmarked subject positions of whiteness.[55] Race could be heralded as irrelevant to biology and placed under erasure when the camera lens confined its frame to the merely biological.[56]

The immodest witness, therefore, was formed through contradictions. Practitioners were simultaneously hailed as representatives of a politicized class (women) and as singular individuals. They were simultaneously implicated agents responsible to that which they studied, and an object of inquiry that spoke for itself. Immodest witnessing was structured by this tension of interplay between women as variegated individuals, as members of a common class, and as participants in a research group collectivity. The assumed common sex of participants excluded other ways of marking difference and drawing together collective life. Moreover, the values of individualized variation and comparison among peers was premised on the bracketing off, and even the erasure of, the complex circumstances that placed some women as agents over the fate of others. The assumption that women were invested in the fate of each other simply by virtue of also being women belied the contradictory ways women were riven and bound by uneven biopolitical topologies of late twentieth-century America.[57] While the figure of the immodest witness performed a radical implosion of the subject/object in observations, making visible the embodied and affective subject in the production of knowledge, the immodest witness simultaneously tended to foreground sexed and individualized embodiment at the expense of a more complexly situated, interlocking, and thus more complicit, map of subject making in knowledge production.

The title of the Boston Women's Health Book Collective's bestselling *Our Bodies, Ourselves* captured the elision between self, woman, and body that permeated the women's health movement and stood in stark contrast to contemporaneous efforts by academic feminists to articulate a "sex/gender" system in which the identity of woman was seen as socially produced and distinct from biological sex. Within feminist self help, the presumption of a corporeal basis to womanhood by some participants further stood in counterpoint to the apprehension of bodies as instances of anatomical individualism (unraced and unclassed) and their ethic of individual autonomy over one's body by other participants. Embodiment was universal, but bodies were individual. Thus, the biopolitical investment in individualized variation practiced by feminist self help posited a particular importance to biological bodies for feminists. Bodies formed a manageable, unfixed corporeal individualism that was not equivalent to the way notions of the academically more familiar sex/gender system cordoned off "sex" as a fixed domain and antithesis of the social malleability of "gender."[58]

In sum, affective entanglements in practices of vaginal self-exam circulated in multiple dimensions. In the moment of immodest witnessing, knowing was an embodied, sensory, and recursive act that situated subjects as particular. In the collective project of feminist self help, affective entanglements formed a moral economy of affirmation—of the happiness of knowing oneself through bonding and of recognition of oneself in others as a politicizable collectivity. At the same time, objectivity was reassembled as a project of self-knowing only possible in politically and affectively charged relations with other subjects.

Immodest witnessing, however, can also be examined through an expanded sense of "affective entanglements." The epistemic value granted individualizing and yet bonding affective knowing was, in the 1970s, attached to a larger political economy that helped to set its conditions of possibility. On the one hand, affective entanglements were a form of counter-conduct reacting to practices of dispassionate, professionalized, patronizing, and even coercive scientific authority. On the other hand, they exemplified tendencies in contemporary American culture that called on subjects to release and express their feelings and desires as political, therapeutic, and entrepreneurial acts. While calls to name and fulfill desires through lifestyle consumption and entrepreneurial vitality were deeply antagonistic to radical feminism, such injunctions were nonetheless part of feminist self help's world, and implicated with their effort to elevate affect as a means to find emancipation.[59] In other words, feminist self help was animated in broader circulations of affect that went beyond their own practices.

Affective Entanglements with Capital and Desire

Taking seriously the "economic," I want to draw attention to how affective economies were also shaped by late twentieth-century capitalism. Over the twentieth century and in the United States, subjects came to be understood as the effects of "machines of desire" and "strategies of desire."[60] While today these phrases might ring as fashionable contemporary slogans of academic theory emphasizing the ceaseless production of consuming subjectivations, they are in fact descriptions of the subject's relation to consumption crafted by some of the most influential public relations and marketing experts from the first half of the twentieth century. For Cold War marketing experts drawing on psychoanalytic notions of the self, people carried with them both conscious forms of rationality and unconscious, irrational "drives," "desires," and "affects" that capitalism could mobilize and give shape to through marketing in order to have something to satisfy commodities with.

While the 1940s and '50s were rife with worries that consumption might find a limit when people's needs were fulfilled, or anxieties that irrational drives could be dangerous threats to the nation, by the 1960s such "affects" were more often understood as dangerous to repress and in need of therapeutic expression. This version from the 1960s of an expressive, affective subject—found in both marketing and counterculture—was called upon to feel and name its uniquely individual desires, and thus become more truly herself or himself in the face of mass culture, even stripping away social constraints to find the authentic kernel of self within.[61] Yet, injunctions to be a subject that finds and names diverse and individualizable desires could help generate new desires, and hence new markets. Entering the age of computers, new practices of marketing demographics emerged called "psychographics" that sought to identity the pattern of desires, "values," and "lifestyles" that people were generating, including within counter-conduct, thereby allowing commodities and markets to mobilize and lure these desiring subjects.[62] New and marginal fashions and political sensibilities were especially generative of novelty, and thus were value producing. An early example of this approach to marketing was the Values and Lifestyles Surveys of the Stanford University's spinoff Stanford Research Institute (SRI; which, in turn, was a famous site of military sponsored research) that offered a schema of "inner-directed" and "outer-directed" personality kinds.[63] In other words, in the 1970s, feminist's affectively charged ways of knowing were coexistent with the marketer's "strategies of desire."

It is significant that in the 1970s feminists began to investigate the function of affective labor—variously encompassed by terms such as "emotional labor" and "social reproduction"—as the unwaged work of caring that women commonly did in families, but which was not explicitly valued with a monetary wage and thus was a form of exploited labor that capitalism depends on, but does not remunerate.[64] Instead of understanding this unwaged emotional work as outside of production, Marxist feminists argued that care was crucial to capitalism because it was necessary for sustaining supplies of labor, that is, the raising of the next generation of workers. Industrial capitalism, Marxist feminists of this period argued, was dependent on patriarchy for keeping social reproduction as unwaged labor while still accruing the benefits of the labor power (as well as consumption) thereby created.

At the same time, those feminists who began looking at the changing shape of women's participation in waged labor within the United States also noted the gendering of even waged emotional labor in the expanding service industries, where women as stewardesses, nurses, waitresses, receptionists, clerks, teachers, child care workers, or customer service representatives had to perform the affective labor of smiling, friendliness, and caring as part of their paid work.[65] It is no irony then that the very first radical feminist speech on consciousness raising was given to an audience of stewardesses. Thus, in the 1970s, a sense of emotional work as a generative, value-producing capacity to feel that businesses could harness and attract was becoming legible.[66]

The affective dimensions of intellectual labor, such as is part of scientific and medical research, was also changing within the landscape of business and technoscience. In emerging hubs of venture capital in Boston and California, passion, creativity, and the ability to sustain social networks of friendships became part of a new affective economy of venture technoscience. Beginning in the 1970s, feminist historians and philosophers of science attended to the place of emotion and care within scientific production.[67] While highlighting the work of emotion in research was at one point seen as suggesting a "women's style" of research, more recently affective labor has come to be seen as central to science and capitalism itself.[68]

What I am arguing here is that the attention to the category of affect—

to emotions, to the sensory, to embodiment—in the 1970s by feminist self help (but also in present-day feminist and queer scholarship) was entangled in economic valuations of affect, *even as affect was promised as a source of potential emancipation and alternative politics.* This is not to dismiss the importance of affect as a crucial dimension of politics and knowledge making and reduce it to merely an effect of economic causes. Instead, I want to suggest a skein of appropriation and reappropriations, of antagonistic and yet enabling relations, of counter-conduct entangled with emerging economic values, that together animated affective entanglements as, on the one hand, not merely economic, and, on the other, not innocently merely epistemic or oppositional.

Ontological Collectivities, Liveliness, and Biovalue

Affective economies might be imagined as relations that are world entangled and world producing. Affective economies are constitutive of feminism, technoscience, and capitalism—even if in antagonistic or contradictory ways. As such, the affective entanglements of feminist self help were productive of particular kinds of ontological politics that did not appear out of thin air to understand bodies as varied, generative, changing, individualized, and responsive. Flexibility, proliferation, variation, sensitivity, and alterability were virtues drawn out into ontological form, yet these attributes, like affect itself, were also noninnocent.

It was in the advanced research groups of feminist self help that the most intensive efforts to rematerialize reproduction as a lush liveliness took place. When the LA Self Help Clinic was still housed in the Women's Center, women would drop by to share their new observations: "Pop, they'd go up on the table and put in a speculum." "'Now look at this, I didn't see this two days ago.'"[69] In self help clinics, participants were encouraged to take daily observations on their own, perhaps including a quick sketch of what they saw in a journal or calendar that they could later puzzle over with the group. These repetitious chronological traces could be assembled into a portrait of minute change over time, further expanding the topography of variation into the dimension of temporal change. Rather than comparing themselves to an abstract, universalized norm (as one might find in a medical textbook), in using the technique of vaginal self-exam they relied on comparisons within small groups of women and with each woman's own changes over time. This schooled attention to slight varia-

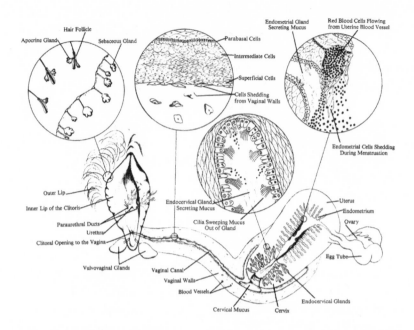

2.6. This unconventional illustration by Suzanne Gage demonstrates how changing the point of entry into anatomical examinations, in this case using a woman's sense of smell, produced alternate ways of describing reproductive anatomy. Here, the sources of smell routed in cellular processes are drawn in magnified bubbles. The cellular processes as a whole make up a "vaginal ecology." FFWHC, *How to Stay out of the Gynecologist's Office* (1981), p. 10.

tions in anatomical detail over time produced a sense of lush, changing variety through which the feminist self help movement sought to remap the anatomical terms of healthfulness. Healthfulness here was a disidentification from abstract norms in favor of shifting dynamic forms.

This rearticulation of healthfulness as a lush garden of shifting anatomical diversity extended beyond macroscopic features to include microorganisms ubiquitously present in vaginas that could cause common and minor, though sometimes recalcitrant, infections. This microscopic variation was dubbed the "ecology of the vagina."[70] (See figure 2.6.) While viewing a drop of vaginal secretions with a microscope, self helpers taught themselves to see "sloughed-off cells from the vaginal wall, a few yeast plants, lots of bacteria and sometimes even a few one-celled animals, trichomonads."[71] If a woman was menstruating, she would see red blood cells. If she had recently had heterosexual intercourse, she might see sperm. As with their gross observations, self helpers were concerned with noticing how the exact constituency of a vaginal ecology would change over time, often in synchrony with the changing pH of the menstrual cycle.

This sense of a vaginal ecology was aided by microscopy. A manual of procedures in a Well Woman clinic, called the *Black Book* (written to defend against accusations of practicing medicine without a license), explained that using a microscope "is simply an aid to better eyesight." "Like wearing glasses, it improves the eyes' ability to detect things. Microscopes can be owned by anyone and are commonly used by students who learn to use them in class. Detecting the different organisms and distinguishing what they are doesn't take any more special ability than does a child's skill in telling what year and make passing cars are."[72] Though strategically represented as a simple magnification of eyesight, the ability to perceive a wet mount slide as a vaginal ecology was by necessity a learned technique of observation that called for assembling details into a relationship of changing diversity. Looking in a microscope was not a neutral gaze taking in a self-evident world; it was a repeated act made sense of by a politicalized apparatus (the self help clinic) that rerepresented entities already codified by conventional gynecology. Strategically characterizing microscopy as simple to defend the legality of lay health care invoked a tension within the ideology of vaginal self-exam, and self help more broadly: at the same time that self help practices were clearly schooled techniques for learning "another way," they were also coded as "commonsense," "simple," "routine," or even at times "instinctive," in order to legitimate the labor of laywomen, both to themselves and to others. According to much of feminist self help rhetoric, an unencumbered understanding of the vagina was available to any woman who overcame social taboos and dared look. The practice of vaginal self-exam not only schooled women in fine-grained sensory observations, it simultaneously refigured their object of study—vaginal ecologies—in a way that authorized the very act of intimate frequent personal observation.

The elaborate daily attention to minute changes of one's vagina and cervix reached its pinnacle in the Menstrual Cycle Study of 1975, also undertaken as part of the Federation book project.[73] Every morning, during a full menstrual cycle, a cadre of nine women gathered to make thirty-six time-consuming observations about their own bodies. "We started out just wanting to measure *everything*, knowing full well—whoa! But let's start with that then we can break it down."[74] Rather than "everything," the kinds of observations made were bounded by a technique of charting minute personalized details. The "woman's point of view" that the study aspired to capture was enacted as a swarm of qualitative and sensory observations. As in a typical vaginal self-exam, the character of the cervical opening, its color, consistency, and amount of secretions were noted. A common set of descriptors were agreed upon and then coded so that observations could be compared in a chart. Observations about consistency, for example, were coded with 0 for unknown, 1 for bloody; 2 for watery; 3 for clumpy or creamy; 4 for slippery, egg whitish, or thinner than a mucus plug; and 5 for mucus plug, clear, gelatinous, or rubbery.[75] The participants also recorded subjective events, such as tenderness, pain, cramping, libido, appetite, headaches, and fluid retention. They even devised an elaborate system for charting their moods from day to day. Though sensory observations dominated, the study also took a handful of simple measurements: saliva, vaginal secretions, and cervical mucus were measured with some pH paper and glucose test-tape; basal body temperature was taken before getting out of bed; Pap smears were done daily; and cervical secretions were gathered with a cotton swab, placed on a slide, dried with a fixative, and then examined under a microscope for their ferning patterns.

Lastly, a daily color photograph attempted to record the visual appearance of the cervix.[76] Many of these photographs were taken by Sylvia Morales, who a few years later in 1979 would make the landmark feminist film *Chicana*. While the film *Chicana* frames women as engaged in diverse forms of labor, the cervical study involved close-ups of cervixes without larger visual context. In the Federation's publication, *A New View of a Woman's Body*, a collection of these photographs was published, with captions adorning each image of a cervix providing details about the woman's age and reproductive history, including births, abortions, sexual activity, birth control method, and surgical procedures. The context provided was entirely biological, with no information about the women's ethnicity or race, class, labor, or other social locators, underscoring variation in a resolutely biological, rather than social or economic, domain.[77] While each image was nearly identical in format, a close-up of a glistening pink circle of flesh, highlighted through the series were subtle differences in shape, tone, or secretion across women and over time. The stage of the cervix was arranged to portray an embodied and variegated individuality.

What was rendered perceptible through this elaborate and tedious collecting and cataloguing of detail? The study did not conclude with a summary description of a menstrual cycle. Nor did it identify a series of markers that identified distinct stages (as discussed in the next chapter). Instead, the study concluded what it was designed to perceive—precisely the converse: women do not match an abstracted cycle, and healthfulness cannot be accurately measured through a once-a-year marker like a Pap smear. Instead of characterizing the menstrual cycle as simple (paralleling their strategy with vaginal self-exam and abortion), they argued that it was more complex than medicine portrayed. What a "normal" menstrual cycle looked like, so they argued, could only be determined by studying "each woman's cycle within the context of the cycle itself as opposed to comparing to a norm," thereby "redefining and individualizing the concept of 'normal' for women."[78] Since doctors relied on annual visits and did not have the time to make such painstaking daily observations on each patient, women occupied a privileged position for understanding this complexity. Using odor as an intimate organizing affective entry point could, for example, produce a disorienting anatomical portrait unlike that found in any medical textbook. Feminist self help strongly encouraged women to use smell, and not just eyesight, to track their vaginal ecology. The woman's point of smell, so to speak, became a way to reimagine female reproductive anatomy in terms of regions of cellular sloughing, portrayed in illustrations almost unrecognizable to eyes schooled only by conventional textbooks.

Materializing reproductive embodiment as a highly valued, dynamic, and generative domain of variation and affect, through practices explicitly considered radical, nonetheless had a synergistic relationship with new modes of valuing living-being, particularly in research with biotechnologies indebted to the reproductive sciences. While feminists of the 1970s were more likely to characterize medicine as treating women's bodies like machines in industrial assembly lines, and while this was certainly a dominant organizational motif for apprehending the body, the 1970s was also a moment of emerging biotechnology, particularly in California, and the beginning of a new era of clinical reproductive medicine concerned with the technical choreography of fertility at micrological levels.[79] Through the arrival of reproductive medicine, the 1970s was an important decade for the refiguration of what the feminist technoscience scholar Catherine Waldby calls *biovalue*. Biovalue names the "yield of vitality produced by the biotechnical reformulation of living processes."[80] Biovalue is constituted out of the generative and remixing capacities of living-being to offer further capacities, processes, and entities not only for the transformation of life, but for the circulation of capital. In ways analogous to the harnessing of proliferate desires as opportunities to reinvigorate capital, so too has generative living-being—and particularly the micrological substrates of reproduction from DNA to cloning to stem cells to tissue culture—become drawn into, though not fully subsumed by, the proliferating incitement of biovalue. Dynamic living difference can trouble life—violating and disidentifying norms and queering ontology—and at the same time it offers a newly intelligible domain for capitalist ventures to capture and incite. Again, this is not an argument that the lush liveliness conjured by feminist self help is reduced to or determined by the inauguration of generative biovalue; instead, I wish to show that they were simultaneous announcements of a new materialization of living-being that could be multiply and contradictorily politicized. As I argued above, calls to express individualized desires not linked to mass culture could enable new modes of hitching affect to capital in marketing. In a similar way, the oppositional ontological politics of heralding the ways living-being can synergistically generate difference and affect reverberate with capitalism's requirement for exuberant territories through which to implant and recycle value.

Topologies of Situated Feminist Technoscience

As Haraway insists with the material-semiotic figure of the Cyborg, I want to track the complicities and the promises of the figure of the immodest witness as a reformulation of objectivity, of feminisms, and of sexed living-being as my own commitment to oppositional tactics of knowledge production.

Relatives, if not the direct offspring, of the immodest witness of vaginal self-exam can be found in more or less altered forms in the scholarship of Marxist-inflected feminisms of scholars such as Dorothy Smith, Nancy Hartsock, Sandra Harding, and Patricia Collins (today retrospectively dubbed standpoint theory), as well as the work of Haraway and Chela Sandoval.[81] Common to these feminist theories are socialist feminist itineraries and visions of knowledge production that begin with located experiences. For Harding's "standpoint theory," "starting thought—theorizing—from women's lives decreases the partiality and distortion in our images of nature and social relations. It creates knowledge—not just opinion—that is socially situated. It is still partial in both senses of the word—interested and incomplete; but it is less distorting than thought originating in the agendas and perspectives of the lives of dominant men."[82] Other feminist standpoint theorists have also increasingly made efforts to map standpoints within the complicated terrain of multiple *locations*, rather than on the basis of an unhistoricized or individualized *corporeality*.

Haraway's theorization of "situated knowledge" crafted another politicized interruption of objectivity, rethinking the standpoint: rather than just something subjugated people brought to science, it became the historically constituted and power-laden conditions of all knowledge production.[83] Haraway's situated knowledge both declares all science already situated in matrixes of capital, subjectivity, and history, and declares itself to be an oppositional project that calls for the mapping of such power-laden matrixes as a necessary part of making better accounts of the world. However, Haraway's work always attends to the noninnocent entanglements of oppositional knowledge making, and in this sense importantly rejects the temptation of romanticizing a purely oppositional knowledge project. As in consciousness raising, the critical potential of situated knowledges must be consciously fashioned, but this still leaves open the practical question of how. How might feminisms and other oppositional projects go about earning this critical perspective? "But *how*," asks Haraway, "to see from below is a problem requiring as much skill with bodies and language, with the mediations of vision, as the 'highest' techno-scientific visualizations."[84] Moreover, Haraway insists, the critical and analytical "Virtual Speculums for a New World Order" that feminists have forged are not for the special case of female bodies, but are applicable to the center of techoscience itself.[85]

As a situated knowledge, then, the broad lesson to be learned from historicizing vaginal self-exam is that the "how" of knowing is as much a question of the promise and limits of affectively charged counter-conduct, as much a question of subjectivation in noninnocent economies, of entangled reassemblies and appropriations, and of marked and unmarked labor as it is a historical episode in the history of objectivity. Rather than an odd marginal practice only of interest to feminists, vaginal self-exam announces a particular reassembly of objectivity in the late twentieth century, with the generative features of counter-conduct, affect, and biovalue at its heart.

Haraway's concept of situated knowledge moved the problem of objectivity away from the subject-object dyad to reformulate it as a problem within political economy, uneven strategies of power, and multiple contradictory subjectivations in noninnocent matrixes. The geographer Cindi Katz has further built on "situated knowledge" to offer the notion of "counter-topographies," which are knowledge-making projects that are not just locally situated. They are also tracked and connected in uneven geopolitical relations of global capital, imperialism, nationalism, and other transnational formations. The notion of counter-topographies is an effort to link multiple situated knowledges across a differentiated world.[86] It is in a closely related sense that this book attempts to historicize feminist self help in Los Angeles on larger uneven *biopolitical topologies*, connecting this book's methods genealogically to vaginal self-exam itself. I have been directly and indirectly influenced by the practices of feminist self help in my scholarly work, as well as a beneficiary from the services of feminist women's health clinics in my lifetime. This chapter, this book, is part of this unfolding trajectory, and I believe there is still more that critical studies of technoscience can learn from feminist practices precisely by attending to the noninnocent historical emergence of "affective entanglements" that still hold so much appeal, and do so much work, today.

Feminist self help, with vaginal self-exam as its iconic protocol, was one of the most sustained efforts to practice science as feminism. It was also a reassembly of objectivity that reverberates in participatory methodologies of many political stripes today. What was the fate of vaginal self-exam? Rarely practiced today, vaginal self-exam declined both for reasons from within and without. As the Reagan era unfolded (and as detailed in other chapters) militant antiabortion activists besieged Feminist Women's Health Centers. The day-to-day harassment and the imminent threat of violence created a "siege mentality" within the centers' walls.[87] The incredible amount of energy, emotional and physical, that went into escorting women through blockades, into clinic security, into court cases, into finding doctors willing to work under the threat of violence, and into rebuilding destroyed clinics drastically redirected the labor of the feminist self help movement.

From within the women's health movement, their very success deflated the consciousness-raising power the vaginal self-exam had enjoyed in the

1970s. A new moral economy of health care arose—calling for the well-educated, well-informed, self-knowing patient to be prepared to advocate for herself as a consumer within corporate medical institutions. Put simply, in the last thirty years the status of white, middle-class women as patients has dramatically changed, and thus so too did the biggest constituency that the affective economy of vaginal self-exam had appealed to. While the practice today is rare, the assembly of affective engagement, particular embodiment, injunctions to empowerment, intimate attention, collective labor, and generative living-being coils forward.

CHAPTER 3

Pap Smears, Cervical Cancer, and Scales

The Pap smear is a lifesaving technology. In the twentieth century, as the scholars Adele Clark and Monica Casper have shown, it became the "right tool for the job" for reducing the deadly scourge of cervical cancer in the United States, Canada, Scandinavia, Britain, and many other countries with well-established health care infrastructures.[1] Compared with the heated controversies over IUDs, hysterectomies, sterilization, in vitro fertilization, and other techniques of reproductive health, the Pap smear was a procedure that brought divergent interests together in relative agreement. Doctors and feminists alike have by and large embraced the Pap smear. As a conventional technology that was not politicized as a problem, the Pap smear provides an entry point into the messy imbrications between feminisms and biomedicine.[2]

More than a moment of swab meeting flesh, the history of the Pap smear draws together labs, labor, public health screening, racialized risk, feminist NGOs, and international health policy—as well as less likely entanglements, from prisons to diamond mines. Even the genealogy of the very term *reproductive health* is attached to the Pap smear. Put simply, the Pap smear was a medical protocol through which feminism, sex, race, economics, transnational policy, and biomedicine collided in the late twentieth century.

This chapter narrates the shifting rearrangements of the Pap smear and its embrace of the deadly problem of cervical cancer over the twentieth century as a history of entanglement and scale. It tracks the politics of the Pap smear as it passed in and out of a variety of feminist projects and health care practices that each changed how cervical cancer and its relation to the Pap smear was understood. Most importantly, the chapter seeks to map these manifold and layered rearrangements of biomedicine and feminism as *spatialized* through particular scales: from the clinic, to the mass screening program, to transnational policy arenas. In these ways, the chapter attends to what this book broadly calls *topologies of entanglement*, that is, the uneven, spatial, and often contradictory traffic of connections that are the conditions of possibility of both technoscience and feminism.

The story of the Pap smear as the solution to the problem of cervical cancer is partially a broad story about the ways health care—and reproductive health more specifically—was recomposed and made newly contestable in the second half of the twentieth century. During that moment, feminists could be found occupying an efflorescence of positions relative to medicine: as denouncers of perilous technologies, as professional participants, and as users and patients. Race also played a crucial role in these rearrangements of the Pap smear, as early twentieth-century raced versions of biological susceptibility to cancer and segregated medicine were transformed into new modes of tracking risk based on nonbiologized statistical racial accounting.

Scale, in this chapter, offers an analytic trope for thinking through the changing ontological politics of cervical cancer, in which its materiality extends beyond the phenomenon itself into labs and even geopolitics. Uneven worlds are folded into the politics of cervical cancer. Attending to scale involves following how health care projects circumscribe and extend the spatialized scope of problems—and thereby remake them. Cervical cells, for example, consist of more than a nucleus and cytoplasm contained by a membrane. Cells, the qualities that adhere in them, the actions that can be performed on them, and the things that can be made with them are concretely rendered knowable and alterable by virtue of the techniques and material conditions for apprehending them, which in turn are shaped by the uneven exercise of power extending along multiple directions in instruments, clinics, professions, laboratory infrastructures, and so on, as well as in broader economic and political arrangements. Myriad relationships can thus extend into the ontology of cervical cancer.

While it is intuitive to think of the Pap smear as a technical solution to the preexistent biological problem of cervical cancer, an inverse account is also possible. How was the problem of cervical cancer posed by the material and social interventions that sought to ameliorate it? In other words, how was the problem of cervical cancer shaped and bound by the solution of the Pap smear? By the problem of cervical cancer, then, I am referring to both a potentially deadly organic condition and historically specific formulations of the causes, nature, and distributions of cervical cancer as well as to the material and technical acts that helped to make cervical cancer knowable, detectable, preventable, and ameliorable. In other words, cervical cancer has a complexly scaled ontological politics that is importantly attached to ways of situating the Pap smear. Hence, the term *scale* indicates more than small and large, or even nested orders; it describes historically specific dimensions, resolutions, relationships, and distributions.

The shorthand *scale* therefore points to the spatial extensions and resolutions at which feminist and biomedical projects apprehended health problems in a world shaped as much by geopolitics as by biopolitics.[3] Feminist projects mapped these scales selectively—as any map does—plotting some features and not others, rendering details at particular resolutions, and marking borders of their problem-space. Scholarship by critical geographers has underlined that scales—such as local, national, and transnational—are not preordained domains. Rather, they are produced by structures and processes maintained by both human and nonhuman actors.[4] At the same time, resolution, points of interest, and edges of a map are not invented out of nothing, but must be assembled and made legible. In this way, feminisms can be understood as fashioning projects that acted on scales already in place (such as scales already generated by biomedicine, the state, or economic development regimes). Yet, when feminists critically acted on these scales they also made maps that highlighted and valued some features and connections, and not others, as changeable and politicizable. For example, feminists variously connected, and did not connect, race to questions of cervical cancer. What differentiates the feminisms I am concerned with here is not primarily that they physically inhabited different parts of the world or were ideologically distinct, but rather that they *differently mapped the extension and nature of the processes shaping the Pap smear and cervical cancer*. Though the feminist interventions I look at emanated from the same general geographic starting point, North America (primarily the United States, but also Canada, and Barbados), I argue they are usefully differentiated by the scales on which they plotted themselves when acting on the problem of cervical cancer and, correspondingly, the question of the Pap smear as a solution.[5] Within contradictory relationships plotted on different maps, the Pap smear could be both the right *and* the wrong tool for the job of the shifting problem of cervical cancer.

The account here selectively tracks only a handful of feminist interventions toward the relatively uncontroversial Pap smear and the problem of cervical cancer. While these feminist projects are chronologically arranged, they do not provide a narrative of linear change over time. Instead, they are layered examples of the ways feminisms mapped the problem of cervical cancer onto three different scales: the clinic, mass screening, and transnational health projects. What was "right" and "wrong" about the Pap smear were far from stable across these differently scaled and mapped feminist interventions.

Offering up stories of the Pap smear both in and out of feminisms, this chapter begins with the initial clinical assembly of the Pap smear in the early twentieth century. At its start, the Pap smear was a crucial element in the history of twentieth-century laboratory medicine, race, and gynecology, with cervical cancer becoming a specific problem of errant cells. In turn, feminist self help in the 1970s developed politicized clinical protocols that reframed the Pap smear, constrained thinking about race, and placed the spotlight of feminist health research on the cervix. The chapter then turns to how the Pap smear was rearranged at the scale of national public health screening in both the United States and Canada, where cervical cancer was remade as a problem of racialized and economic risk. Feminists of the 1980s, in response, variously remapped the problem of cervical cancer as an effect of political economy, in which racism, screening, and health care were complicit. In the final section, the Pap smear moves to the scale of transnational reproductive health politics in the 1990s, where American and Barbadian feminist organizations declared the Pap smear the wrong tool for the job for detecting cervical cancer, now understood as a sexually transmitted infection, thereby entangling the Pap smear with questions of postcolonialism, the NGO-ization of feminism, and family planning, helping to craft the very term *reproductive health* as an object of international population policy. I leave for the chapter's conclusion speculations on the emerging rearrangements and possible decline of the Pap smear with the recent circulation of a vaccine for a global market.

In this way, the story of the Pap smear both moves up scales from the politics of cells to transnational health policy, as well as shifting over time from industrial laboratory medicine in an era of racial desegregation, to a time when public health became concerned with the regulation of racial and economic risk, to a more contemporary moment shaped by debates about the political limits of American protocol feminism in a postcolonial era.

Assembling, Professionalizing, and Making Cellular

The Pap smear has been for over half a century a normalized feature of reproductive health care. Dating back to the 1940s, Pap smear screening has been the longest-running, most-widespread cancer screening program in the United States and Canada, making the cervix one of the most intensely scrutinized body parts. The Pap smear has also been tremendously successful. Before screening, cervical cancer was the most common cause of death by cancer for women in North America. By the end of the century, cervical cancer mortality in the United States had been reduced by as much as 70 percent.[6]

Pap smears and cervical cancer, moreover, were crucial to the emergence of professional gynecology. Yet, in nineteenth-century America, during the dawn of the profession, cervical cancer tumors, recognized for centuries because of their accessibility to visual examination, were a highly stigmatized disease. Treating "uterine cancer," which at that point could mean either both uterine cancer or cervical cancer, was then controversial; the board of the first hospital for women, the New York Women's Hospital, wanted to refuse treatment to impoverished and often immigrant patients with uterine cancer because of the disease's smell and stigma.[7] As physicians developed surgical methods in the late nineteenth century—establishing the specialty of gynecology as a distinct field—cervical tumors, cervixes, and entire uteruses increasingly were removed surgically. As cervical cancer was typically asymptomatic in its early stages, yet often fatal when more advanced, surgery offered the first promising avenue of treatment.

Early gynecology in the United States, as a largely surgical practice, was from its inception stratified, for example, at once exploiting enslaved and poor women as experimental subjects, racializing many gynecological problems as artifacts of white civilization, and embracing sensitive elite white woman as its normative patient.[8] Cancer itself tended to be racialized as a "white" disease in the early twentieth century, while through the visibility of breast and uterine cancers it was often gendered female.[9] Early surveys of the incidence of cancer, such as by Frederick Hoffman, explicitly associated cancer with white "civilization."[10] Just as some physicians argued that blacks had a natural predilection to some diseases, such as sickle cell anemia and tuberculosis, so too did physicians claim that "Negros," "natives," or "primitives" were more immune to other illnesses associated with high levels of sensitivity and civilization, such as ovarian tumors or cancer more broadly.[11] Early American Society for the Control of Cancer representations widely portrayed the disease as afflicting well-to-do white women, urging them not to be silenced by fear.[12] In the years of Jim Crow America, white female patients were actively brought into the fold of cancer detection.[13] This history of racializing cancer would later be incorporated into Pap smear screening.

The way cancer was racialized in the course of the twentieth century had as much or more to do with segregation of care and continued practices of virulent racism within medicine as it did with the racialization of specific diseases. In the United States, not only was access to health care configured by a patchwork of state segregation regimes; the ability to pay for medical care in a nation that did not establish a national health care system had profound class implications for when, if, and for whom professional health care could be afforded at all. Even when cervical cancer was not itself explicitly raced as an outcome of biological difference, medical knowledge about cervical cancer in the first half of the twentieth century was nonetheless crafted on a racialized political economy of medicine. Thus, physicians typically took as given the whiteness of prospective paying patients, while public health projects, in an era of eugenics, targeted variously racialized and stigmatized subjects as potential threats to the health of the nation. In other words, even the ordinary workings of medicine, science, and the state that did not take "race" as an explicit object of inquiry were nonetheless configured through race in the intensely and complexly racialized conditions of early twentieth-century America.[14]

Whereas gynecologists, who were in the process of professionalizing at the end of the nineteenth century, increasingly approached advanced cervical cancer as a surgical problem of organs, pathologists armed with microscopes were beginning to redefine cancer as a phenomenon caused by errant cell growth. Cancer became a new kind of problem along with its definition as a cellular condition. Moreover, the Pap smear was instrumental in this shift, becoming a prototype technique for envisioning and diagnosing cancer through cells. With the help of a Pap smear, cancer was imbued with new cellular time scales of growth, "death," and regeneration of cells more generally as micrological living entities.[15] Through cells, cancer could be identified at earlier, less fatal stages. When cancer was understood as a cellular phenomenon, it became imaginable that pathologists could detect cancer earlier through biopsies or other practices that put cells under the microscope.

The technique of the Pap smear was first assembled at a small laboratory at Cornell Medical School in New York. In 1917 George Papanicolaou, the Pap smear's namesake, announced the use of vaginal smears as artifacts for reading stages into the estrus cycles of rodents. Thus, as Adele Clarke has eloquently shown, initially the smear was not a cancer-detecting device at all but a means for researchers to quickly read the stage of estrus in living experimental animals, such as guinea pigs, used in the emerging field of reproductive science.[16] The Pap smear allowed reproductive scientists to quickly assess the stage of an animal's estrus cycle and correlate it with other measures, particularly of hormonal fluctuations. It was a simple method that spread rapidly through the field, and at the same time sped up the pace of reproductive science research itself.

In 1933 Papanicolaou and his collaborators succeeded in creating a system of reading smears applicable to humans—an alphabet of cells associated with phases in the human menstrual cycle.[17] The presence of cellular states on a slide became a code that was matched to a cycle of tissue changes in ovaries and to hormone levels (see figure 3.1). Thus, the Pap smear as a technique both materialized and coded a new object of inquiry, a "typical" human female reproductive cycle manifest at the cellular level.[18] The "implicated actor" of the Pap smear—the subject for whom the design of the technology was meant and imagined yet who was absent from research and design—was women in general and abstractly.[19] Technically, Pap smear interpretation required a standard of an abstract "normal" set of cellular states, a physiological standard not determined by extensive epidemiological study but initially crafted out of George Papanicolaou's spouse Mary's bodily specificity, then elevated into abstract universality.

Tracking this "normal" cycle of cellular change, in turn, demanded identifying and sorting out abnormal cells. In 1941 (followed by a detailed monograph in 1943) Papanicolaou published a classification system for reading and grading cells from normal, to early cancer, to invasive can-

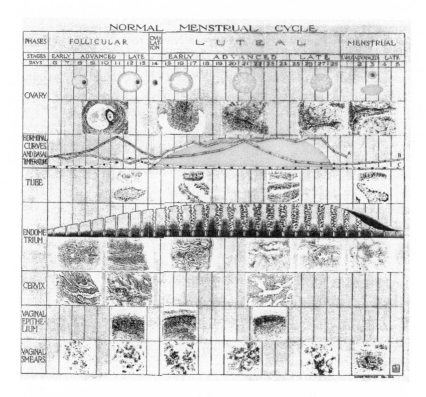

3.1. Illustration that charts how the reproductive cycle of the human female is correlated to changes over time in vaginal smears (bottom row), tissues, hormone levels, and the ovary. From George Papanicolaou, "The Sexual Cycle in the Human Female as Revealed by Vaginal Smears" (1933).

cerous, garnering wide attention.[20] Thus, cancer at the cellular level was plotted as a temporal progression. The Pap smear became a tool through which cells could be read along two temporal axes: first, as a representation of an abstract menstrual cycle and, second, as a representation of the emergence and progression of cancer. Since surgical techniques were quite successful for treating early cervical cancers, the possibilities for the Pap smear were profoundly lifesaving.

A conventional Pap smear was performed as follows. Exfoliated cells of the cervix were removed from the vagina with a pipette, swab, or wooden scrapper and then spread on a glass slide, thereby taking cells outside of the body and turning them into a representation that could be viewed under a microscope. A stain was then added to the smear to bring into view specific cellular qualities that were then dividable into stages or phases (see figure 3.2). Third, a sign system was used for reading into the smears the temporal stages of organs, hormones cycles, and possible cancers still in the body. Thus, in a Pap smear, cervical cancer could be screened as a problem of cells "readable" with the aid of microscope, stain, and alphabet.

As Clark and Casper have shown in detail, over the 1940s a concerted effort was necessary to make the Pap smear the "right tool for the job."[21] Papanicolaou called the vaginal smear a method "so simple and inexpensive that it may be applied to large numbers of women."[22] But even Papanicolaou qualified his excitement by noting that reading smears required a "careful and discriminating cytologist who has had experience in the field," rather than occasional readings of individual smears by physicians.[23] Thus, for the technique to be used on large numbers of women—as a mass detection technique—it required simultaneously generating large numbers of smears and building a laboratory infrastructure with skilled cytologists to quickly read them.

In 1945, the American Society for the Control of Cancer, founded initially around uterine cancer, changed its name to the American Cancer Society (ACS) with the slogan "every Doctor's Office a Cancer Detection Center."[24] Papanicolaou's vaginal smear fit well with the ACS's agenda and the organization began, along with the newly formed National Cancer Institute, to throw its weight behind the smear, even funding its medical and scientific director to travel the country teaching pathologists how to read smears.[25] Though there was no epidemiological data to support any specific recommendations regarding the frequency of Pap smears, and some pathologists even objected to its use-value, the American Cancer Society and the American College of Obstetrics and Gynecologists both recommended annual Pap smears for sexually active women and/or women over 18.[26] The Pap smear offered hope that medicine could intervene in the deadly incidence of cancer.

The Pap smear, now crafted into a preventative screening program, was pivotal in the reorientation of gynecology in the 1940s and 1950s from a profession dominated by surgery toward one of "prevention," counseling, and other "well woman" procedures.[27] Since the Pap smear could only detect, and not prevent or treat, it became a crucial impetus for the yearly gynecological examination of the not-yet-sick patient, constituting the cervix, and by extension female reproductive organs more broadly, as a domain of what scholar Kathryn Morgan calls "virtual pathology," that

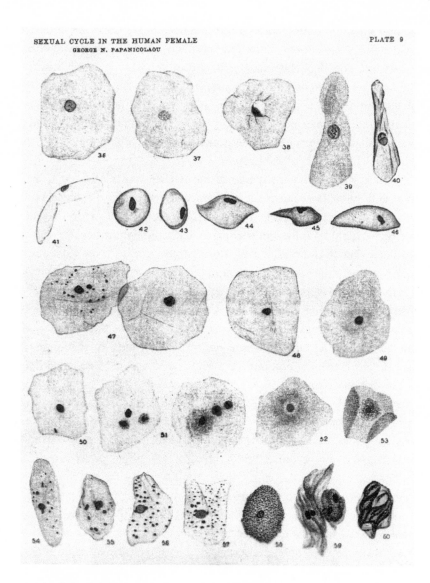

3.2. A cellular alphabet created by George Papanicolaou to "read" exfoliated cells of the human vagina for both the menstrual cycle and abnormal or cancerous cells. From "The Sexual Cycle in the Human Female as Revealed by Vaginal Smears" (1933), p. 635.

is, as a site of potential illness and risk in need of monitoring.[28] Thus, by mid-century the patient-figure of gynecology was not just the neurotic elite white woman of the nineteenth century; she was the robust, health-ful figure of the predominantly white, well woman, hailed as potentially ill but, importantly well and, not-yet-ill. This subject-figure, who might be at risk and whose healthy, marriageable heterosexuality needed preser-vation, slightly differed from the already "at-risk" patient, announced as abnormal before illness was present.

Within this double emergence of the at-risk woman, already abnor-mal, and the well woman as subject-figures of gynecology, Papanicolaou's classification system of cells was replaced in 1956 with a World Health Organization (WHO) System that expanded screening from early cancer to a new type of technical entity—"precancerous cells."[29] The emergence of the precancerous cell as an object of screening was, moreover, not only established as a world standard; it was accompanied by the actual estab-lishment of regional public health screening programs. Thus the recipi-ents of the Pap smear expanded from the paying client to the "mass" of all women and a biopolitical problem of public health. In 1949, one of the earliest screening programs was set up at Vancouver General Hospi-tal in British Columbia, Canada. Other early North American programs were in Ohio, Tennessee, and Kentucky. The Pap smear became a crowning achievement in the American Cancer Society's efforts to promote routin-ized cancer detection, inspiring the promotion of breast examination and other screening procedures.[30]

The success of the Pap smear also consolidated the field of clinical cy-tology, which sought to develop other systems through which pathology could be detected through cellular change. As one of the earliest mass screening projects that depended on laboratory analysis, Pap smears, with their sheer quantity of readings that the widespread annual exam gener-ated, created a dense infrastructure of industrialized laboratory facilities. Thus, the Pap smear was a fundamental component in the rearrangement of medical practice as a risk-based, industrialized, laboratory-centered mode of diagnosis.

Through these rearrangements, the technique of the Pap smear en-tangled together potential cellular pathology, and the heralding of differ-entially risky women, with the industrialization of medical clinical labo-ratories. The Pap smear contributed to a popular apprehension of bodies as populated by cellular entities that needed to be monitored and treated.

It called on women as subjects charged with monitoring their health as a site of risk in the form of their cellularity. At least since the 1950s, posters and postcards have reminded "every woman" to get her yearly Pap smear.

Gynecologists themselves commented on this "progress" and expansion of their duties through the annual exam in their professional journals. In the most expansive terms, one Milwaukee gynecologist declared, "Every gynecologic examination must be the examination and evaluation of the entire woman. The complete office gynecologic examination must include a systematic evaluation of all areas from head to toes. It is our purpose and obligation to the patient to discover, if possible, any or all existing pathologic conditions, whether related to gynecology or not."[31] It was not uncommon for gynecologists in the 1950s and 1960s to also recommend expanding gynecology to include "counseling," ritualizing new patient encounters around the "premarital exam" and "preconception exam," such that "in a limited way, the annual cancer detection examination can turn out to be a preconception procedure for many white women in the child-bearing period."[32] The Pap smear and the annual pelvic exam would then become strategically coupled with another reason women went to gynecologists: the availability, through doctors only, of new mass-produced forms of contraception. Ideally, according to one "progressive" gynecologist from Ohio who wrote repeated promotions of the annual exam, by focusing gynecology on the "so-called well woman" a new dictum began to shape the profession: "the patient is now never discharged."[33]

A 1958 survey found that all of the 750 gynecologists asked had begun promoting annual Pap smears and "well-woman" exams.[34] Over the 1960s, family-planning clinics, serving racially and economically diverse constituencies of women, also incorporated Pap smears as a routine aspect of "basic health" (such as blood pressure) provided along with the techniques of reproductive management. In 1974, approximately 56 million Pap smears were performed that year in the United States.[35] Combined with the increased routinization of the hospital as the most frequent site for births in the 1950s, and the 6.5 million women who began taking oral contraception in the United States between 1960 and 1965, gynecology, obstetrics, and family planning together became the most pervasive points of entry into biomedicine for healthy women.[36]

The cohort of mostly white women in the United States and Canada, who founded the first feminist health centers and collectives in the early 1970s, therefore, was of the first generation whose reproductive health experiences occurred within this reconfigured and expanded "preventative" gynecology. When radical feminist activists in California began organizing some of the first American feminist health clinics in the early 1970s, they were members of a generation who had not only lived their adult lives under the call of the yearly gynecological exam but who also tended to be members of those demographic constituencies who had benefited from the racialized distribution of affluence in the post–Second World War years.[37] For that cohort of women privileged enough to be able to choose to participate in preventative medicine, rather than be targeted by an often coercive racialized public health logic, the patronizing authority of the gynecologist loomed large as a problem.

Feminist health politics would be shaped by this selective call to make oneself available to biomedicine, for which the Pap smear was an important component. The multiple subject-figures of the Pap smear—the privileged well woman securing her individualized healthful vitality and the at-risk women of mass screening for whom cervical cancer was calculated as a deadly statistical possibility—were the effect of the very different conditions in which Pap smears could be situated. In other words, the Pap smear brought together injunctions to ensure risk-free healthfulness joined with population-level calculations of death and birth prevention shaped by profoundly raced and classed logics. Within stratified biomedicine, the call for every woman to make herself available to an annual Pap smear was thus multivocal. Opportunities to have a Pap smear through family planning could be hitched to the logics of population control, which tended to target women who did not or could not seek out the regular care of gynecologists. Moreover, following the legal desegregation of health care, doctors encountered patients who exceeded the norms of white citizenship through public hospitals, emergency rooms, and "community" health projects that simultaneously expanded access to public health care and brought diverse women under medical scrutiny through the rubric of "family planning"—thus offering the Pap smear to more women. The Pap smear was simultaneously lifesaving, instrumental to the arrangement of professional gynecology, and incidentally proximate to sometimes coercive, racist, and unequal medical practices. Or put another way, the tensions between preventative gynecology, family planning, surgical sterilization, reproductive sciences, cytology, obstetrics, cancer screening, and state public health all coil through the history of the Pap smear.

Feminists Reassemble the Clinic

The radical feminist health collectives up the West Coast of the United States and into Canada of the 1970s (which are the starting points of this book) explicitly reconfigured and politicized the clinical moment of the gynecological exam, and hence the terms by which women encountered the Pap smear. Along with the speculum, the Pap smear was a tool that they sought to "seize" as part of a biopolitical project to "take control of our own bodies." At the scale of the clinical exam, not only was the Pap smear appropriated, but the ontological politics of cervical cancer, along with the precancerous cell, was reconfigured as feminists offered ecological ways of understanding cellularity. These interventions into the Pap smear, however, were largely a secondary consequence of their more general protocol project of reconfiguring the gynecological "well-woman" exam as a self help and participatory practice.[38]

As a protocol feminism — that is, as a politicization at the level of techniques — feminist self help tended to map clinical practices as their scale of intervention, the resolution where they could appropriate practices and create a domain of "woman-controlled" health care.[39] For example, "peers" and a lay health-care worker, not a doctor–patient dyad, were the core participants in a feminist self help clinical encounter. Feminist protocols rearranged the practices through which people addressed one another (with equal status), what they wore (nonprofessional clothes), what the clinical setting looked like (homelike), the distribution of tasks (patients were as active as health care workers), and so on. Thus, feminist self help mapped the protocols governing the clinical encounter as an intensely politicized terrain of intervention. Vaginal self-exam was in many ways the practice at the ideological center of feminist self help. It made the *cervix*, lit in the spotlight of the hand-held flashlight, the star attraction in their restaging of the gynecological exam, far more so than professional gynecology ever emphasized.[40] As a component of such group efforts to put the cervix in the spotlight, participants aided each other in performing Pap smears.

Papanicolaou had broached the possibility of collecting a Pap smear from oneself as early as 1943, and while he advised that self-collecting a smear did not affect its quality, the profession of gynecology was reluctant to give authority over the smear to women themselves, instead absorbing the smear as a key practice within a reorientation of gynecology that required an annual pelvic exam and brought increasing numbers of women under its care. While it is technically possible to perform a Pap smear on oneself given the correct equipment, a second person could more adroitly manipulate the typical wooden spatula or cotton swab used in the 1970s. For feminist self help, taking each other's Pap smear was, like using a plastic speculum, a symbolic and material moment of taking medicine and technology into "one's own hands," a remaking of the patient and caregiver into the subject-figure of the sovereign woman. Appropriating the term *well woman* from prevention-oriented gynecology, feminist self help embraced the figure of the already healthful woman charged with maintaining a risk-reduced vitality. Yet, in feminist self help, the well woman not only made herself available to medical scrutiny; she enacted it.

In addition to mutually assisting in Pap smears at participatory clinics, some feminist centers even encouraged women to observe their own smears with the microscope. However, since correctly reading a smear required training and practice, most feminist health centers sent smears to conventional labs. Marking a limit to the ideal of creating a self-sufficient feminist health practice, the Pap smear was not an important technology in their iconography. The Pap smear, instead, can be historicized as expressing some of the core contradictions within feminist self help, which aspired in its most radical form to escape from biomedicine, but nonetheless drew on the wide availability of medical commodities and the financial flows of Medicaid in clinical practice. The movement of the Pap smear from participatory clinic to a lab drew the limits of feminist self help's scale of action. While the Pap smear would physically be analyzed in a laboratory, the results could still be critically reinterpreted back at the feminist health clinic (see figure 3.3 a, b, and c). Within a participatory exam, for example, the health care worker at a feminist clinic would ideally take the opportunity of a Pap smear to start a conversation about the ways laboratories reported results, so that "women can interpret results for themselves."[41]

This protocol politics hoped to dramatically create feminist health care as a space where medicine and technoscience worked differently, an ambition envisioned and accomplished within a very specific, separated, and localized terrain—the exam. Their commitment to antiauthoritarian practices and the insistence on a "woman-controlled," "information-sharing" environment, implied that the category of woman was a crucial and suf-

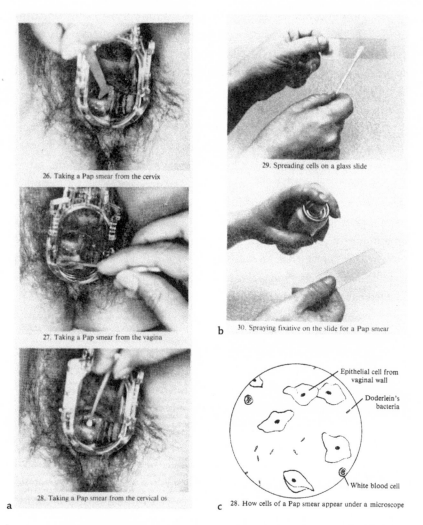

26. Taking a Pap smear from the cervix

27. Taking a Pap smear from the vagina

28. Taking a Pap smear from the cervical os

a

29. Spreading cells on a glass slide

b   30. Spraying fixative on the slide for a Pap smear

Epithelial cell from
vaginal wall

Doderlein's
bacteria

White blood cell

c   28. How cells of a Pap smear appear under a microscope

3.3 a, b, c. Steps for undertaking a Pap smear, published by the FFWHC. Notice how the steps concentrate on the moment of collection, and then skip to a drawing of a smear under a microscope, leaving out the progression to a laboratory. From FFWHC, *How to Stay out of the Gynecologist's Office* (1981).

ficient point of common identity for constituting peers and creating conditions for the even distribution of power in a clinic. In this feminist reconfiguration, the scale of the clinical moment was marked as an unraced space, as a domain where antiauthoritarian practices and solidarity across sex would produce self-determining subjects that did not need to consider race in the moment of politicizing practices.[42]

While the antiauthoritarian and antiprofessional practices of feminist self help was in many ways a radical intervention, it can also be seen as enacting and intensifying a broad historical shift in the moral economy of the patient. What counted as good patient behavior, for example, was shifting from passive obedience to actively choosing, and thus such patients needed to be counseled and informed. "Informed consent" was an ethic-in-the-making in the 1960s and '70s that feminists protocols of information sharing helped to foster.[43] Doctors and researchers struggled to find forms, labels, or visual media that could efficiently standardize informed consent. In turn, through emphasizing information sharing, unraced feminist self help clinics sought to craft practices that democratized and individualized informed consent.

By seeking not only to value the layperson's contribution to health but also to transfer the *work* and *responsibility* of health care and knowledge production to themselves — universalized as "women" — feminist self help advocated and exemplified a wider cultural shift around what behaviors defined the good patient and how health was best self-governed. Rather than simply compliant and obedient, the good patient, over the course of the late twentieth century, became someone who was educated enough to ensure doctors had negotiated "informed consent," and who could be her or his own advocate, as well as someone who regulated her or his own risk and "lifestyle" for the sake of good health. The good patient was someone who made herself/himself available to a doctor as a risk-aware subject before pathology occurred, fashioning a new health moralism of the liberal middle class.[44] The good patient was someone who acceded to health-seeking behavior, was rational and not "superstitious," and knew when to consult a physician and when not to. The "operability" of the good patient — that is, the willingness to make oneself available to biomedical interventions — was a value within feminist self help informed by a calculus of liberal self-determination.[45] Health seeking in the late twentieth century became a moral imperative that responsibilized individuals through risk at the same time that it became a feminist site. Feminist self help, then, rearranged the Pap smear in a way that participated in this imperative, but through a "radical" unraced responsibilizing liberal mode that was deeply antiauthoritarian.

Feminist self help's unquestioned use of the Pap smear in the 1970s was further animated by the new pervasiveness of screening projects—from sickle cell anemia to syphilis tests—that operated through this dual figure of the good patient and the at-risk patient.[46] In their feminist repetition of the imperative to screen for cervical cancer, a single yearly Pap smear was inadequate. Potential abnormality, in their view, was best judged in terms of what was normal for each individual woman, thus critiquing the standardization—the abstract normal—that came with the mass production of preventative medicine in industrial laboratories. Their practice of apprehending bodies as individually variegated cases of normality, in which differences between women were characterized on individual physiological and microbiological levels, combined a sense of the individual uniqueness of each face with the dynamics and balance of an ecosystem of microbes, cells, and fluids.[47] Normality could only be determined at this individual level and over time, afforded by repeated and routine visual inspection. This ontological politics of variegated individuality drew special attention to two of the biological entities already foregrounded by the Pap smear: the cervix and exfoliated cells.[48] Rather than positing the repeated scrutiny of the cervix as a search for pathology, they emphasized the beauty and lushness of the cervix and the fragrant abundance of normal exfoliated cells that varied for each woman in a shifting ecology.[49] The cytological alphabet of cellularity through which Pap smears were read was conceptually turned into a dynamic "vaginal ecology." Precancerous cells became an aberration of the vaginal ecology, which in turn needed to be cared for, appreciated, and monitored at an intimate scale. With careful attention, precancerous abnormalities could recede before turning into cancer. Feminist self help called for *more* intensive surveillance of the cervix than recommended by gynecology.

Thus, while feminist self help employed the gynecological term *well woman* to describe the subject of their own health endeavors, they emphasized varieties of wellness, revealed by a more continuous and intimate surveillance practice. Reproductive health problems generally, and by implication cervical cancer specifically, became problems best detected and prevented when women themselves attended to the practices of detecting and treating them.

In these ways, feminist self help clinics were eager participants in Pap smear surveillance. In the traveling road show conducted by the early members of Los Angeles–based feminist self help, they unequivocally recommended that "one a year is a good idea."[50] Moreover, they recommended frequent vaginal self-exam as an additional potential guard against cervical cancer, as "we can see when there is a major change from what we know to be normal for us."[51] Surveillance was to be internalized, routine, and never ending. Not just a war on cancer, nor a war against precancer, but an ecology to be tended, feminist self help extended the temporal scope of screening in such a way that everyone was subsumed in the need and labor of intimate surveillance for which one's very sovereignty was at stake. While preventative gynecologist talked about the ideal of the "patient who is never discharged," feminist self help advocates encouraged continuous self-monitoring and self-care, turning health-regulating behavior into a ongoing, politicized, self-making, ecology-charting endeavor, never fully complete.

This feminist unraced politicization of the Pap smear, then, was less explicitly oriented toward the problem of cervical cancer and instead highlighted the reassembly of clinical encounters and plotting of precancer onto a vision of vaginal ecologies and well women. Rearranging the Pap smear through a focus on the pelvic exam, feminist self help mapped the clinical encounter as a rich problem-space. At the same time, Pap smears still traveled out of clinical encounters and into industrialized labs as an unproblematized condition of possibility. Such entanglements made it all the more pertinent that as the 1970s came to a close, this imbrication between the scale of the clinic and of industrialized medical infrastructures would become increasingly politicized. The Pap smear exceeded the bounds of the clinical moment, not only depending on labs, but also spreading into a public health project of national importance.

Mapping Screening, Racialized Risk, and Capital

By the end of the 1970s, Pap smears were a widespread and routine technology encountered by the vast majority of women in the United States and Canada. Screening extended individual encounters into a national problem of public health and the distributions of risk in a population. Moreover, the practices of public health screening, as the 1980s unwound, were reinvigorated through new neoliberal and racialized governmentalities. While not a state-mandated test in the United States, the Pap smear had become a form of cancer screening that the state subsidized, promoted, and monitored. The term *screening* captures the dual biopolitical logic of the Pap smear: an act to shield or protect, as well as an act of sifting and detecting disease that is not itself a treatment. The Pap smear both promised a protection of life and at the same time a detection of threats to life. As a screening program, it was covered federally in the United States by Medicaid, but not necessarily paid for by private medical insurance programs, though about half of the state legislatures passed laws to assure its coverage. Thus, in the United States, the Pap smear was a feature of the state apparatus and a commodity.[52]

While there was general consensus in the medical community that screening with the Pap smear was dramatically reducing deaths by cervical cancer, disputes nonetheless percolated over the management of screening as a dual formation of capital and public health. Screening was thus articulated as a problem of accuracy, expense, and overscreening that came with processing the vast majority of adult women, who produced mostly negative (that is, healthy) smears. By the 1980s, sensational news stories and feminist exposés revealed how the accuracy of Pap smear readings was impaired by conditions in industrialized laboratories.[53] Labs hired trained cytotechnicians, positions occupied almost entirely by women who had carved a place for themselves in the biomedical assembly line.[54] Cytotechnicians were under tremendous pressure from both cost-saving logics of the state and profit logics of business to process smears more quickly and cheaply. Like other sites of industrialized labor, laboratories had sought ways to reduce the expense of their operations, quickening production through strategies such as paying by the piece, setting quotas, or even organizing smear readings as a kind of industrial homework to reduce overhead. The worst instances, in which rushed piecework led to missed cases of cervical cancer, were dubbed "Pap mills" by the press.[55]

In response to the scandal of Pap mills and other faults in screening, biomedical and public health professionals reevaluated the Pap smear. The central question, however, was not so much the *usefulness* of the Pap smear in detecting early cervical cancer, but instead its *efficiency* as a screening program. This switch in orientation expressed profound transformations in the logic of health care provision, ushered in at the conjuncture of civil rights and neoliberalization. First, from the 1960s forward, civil rights legislation in the United States demanded the desegregation of medical associations, hospitals, and clinics while at the same time Medicaid in the United States (and universal health care in Canada) induced a dramatic expansion of health care provision beyond the paying classes.[56] The election of Ronald Reagan as president in 1980 marked the crystallization of a second kind of reorganization in the logic of health care as efficiency was calibrated through cost-benefit logics within an increasingly privatized and corporatized "industry." More and more, health care was coordinated through insurance and HMOs (Health Maintenance Organizations), which in turn were governed through risk-based calculi of profits and other modalities of efficiency.[57] The Pap smear was manifest of both these trajectories—a screening program for everyone, yet subjected to neoliberal cost calculi.

The politics of efficiency in both public and private health rearranged screening as a raced mode of accounting. On the one hand, surveillance data were collected nationally by the CDC (or in Canada by Health Canada) to track incidents of cancer and precancer, as well as the prevalence of Pap tests amongst women, with the goal of screening all women. These data, moreover, were sorted by categories of race in order to calculate differential risks to cervical cancer. At this scale of mass screening and racial accounting, *not* having a Pap test became the pivotal risk-factor of cervical cancer. On the other hand, professional, state, and laboratory studies each gathered data on the economic efficiency and cost-benefits of the Pap smear in relations to lives saved. For example, the cost per life saved by treating more curable early cancer in situ was comparable to the cost per life saved of treating more deadly, invasive cancer later.[58] In this way, health accounting could become economized calculations of the value of preserving life.

Critics of the Pap smear noted that the expense of current screening programs was not warranted since screening tended to reach the "wrong women," the typical middle-class and low-risk repeat patients of preventative gynecology.[59] The Canadian province of British Columbia, which not only put into place one of the earliest screening programs in 1949, but also collected rare longitudinal epidemiological data from that time forward, noted that though their program was reaching 75 percent of the population, it was nonetheless "screening the better educated and more informed women." The study explained that "the type of woman who would take advantage of a screening programme is the same one who would present herself to a doctor with very early symptoms from clinical disease. Conversely, the woman who would ignore symptoms as long as possible and seek medical attention only when unable to carry on is, in general, the patient who would ignore such a screening program."[60] Some commentators brought the question of cost-benefit down to the patient's own pocketbook, asking "is the consumer, i.e., the screened woman, receiving good value for her money?"[61] The "right" women, where value was to be found in the form of likely cervical cancers prevented, were increasingly identified as those missed by screening, the data for which were sorted by race.

In Canada, where every province had established a universal health care plan, the government commissioned an influential task force to investigate the efficiency of cervical cancer screening, culminating in the internationally cited Walton Report of 1976 (which would influence policies in the United States). It suggested a new algorithm for testing based on aged-based risk: efficiency could be improved if low-risk women received Pap tests every other year, and every three years if earlier tests were negative until thirty-five years of age, after which screening every five years was sufficient.[62] Despite the Canadian Society of Gynecologists and Obstetricians' rebuttal that the annual exam and Pap smear was "the mainstay" of their profession, the Canadian standard of care was altered in response to the new recommendations.[63] In the United States, where private medicine ruled, the American Cancer Society and professional gynecology successfully rebutted the Walton Report and maintained the annual exam as the standard of practice. The scale of screening was thus producing its own uneven effects.

The logic of efficiency in screening, moreover, viewed the desired subject of the Pap smear through calculi of racialized risk. Here, state-collected medical surveillance data that demarcated differences between women according to categories of racial identity at once captured the work of racialization in social stratigraphies that had shaped women's uneven access to health care, the work of the state in accounting for its citizens by race, and the rearticulation of race as a risk factor. "Race" as a category was necessary, epidemiologists increasingly argued, because it named the material outcomes of political economies, not because of any claims to biological susceptibility or immunity.[64] At the same time, such categorizations of race made possible new ways of governing health. The Pap smear, then, should also be mapped onto the history of public health as a formation of the *racial state*, a term the scholar David Theo Goldberg has developed to describe how the techniques and apparatuses of the modern state have served variously to fashion and modify racial expressions, inclusions, exclusions, and subjugations (for example, in eugenics).[65] Applied to the 1980s, the concept of the racial state is better reframed as a broader *racial governmentality*, that is, as a technique of accountability for governing life through a reproduction of "race" as a debiologized proxy category of political economy, culture, risk, and differential value.[66] What this term highlights is how "race" is doubly at work in racial governmentality—as the material effects of past racisms and as well as a epistemological remaking of race in the present.

The sociologist Troy Duster calls this double work a "feedback loop of knowledge production" about race in biomedicine.[67] In the 1980s, race had become associated with cervical cancer, not because of claims of inherent biological difference, but because of racialized social differences or behaviors coded as culturally, rather than biologically, laden. Race was treated not as a biological kind but as a proxy of cultural difference, helping to give birth to a form of "cultural racism," in which degrees of rationality, backwardness, and pathology were assigned to cultural practices named through anthropological and civilizational logics as raced or ethnic.[68]

In this reinvigoration of public health through racial governmentality, then, not only was cervical cancer data collected and organized through categories of race, but by the 1980s an abundance of public health and medical literature emphasized that "beliefs" belonging to ethnic or raced communities were a significant explanatory factor in the lack of health care.[69] Mexican Americans, for example, were said to have a cultural "fatalism" about cancer that stopped them from seeking screening.[70] Black Americans were "spiritual" rather than rational about their cancer.[71] Against expectations in the 1980s about racialized risk, Inuit women were found to have no cervical cancer, despite both being "native" and having a "cultural acceptance of early sexual intercourse."[72] This cultural racism layered upon the early twentieth-century view that different cancer mortality rates were related to different levels of civilization, where "primitive" noncivilized peoples around the world, including slaves and Indians, purportedly did not get cancer because, researchers speculated, of their primitive cooking methods, diet, and simpler lives.[73] In the late twentieth century, then, *race* became shorthand for descriptions of the behaviors and conditions through which people lived. In so doing, the concept helped to reposition the Pap smear as not only a tool of cancer detection, but as an unevenly distributed practice expressible in terms of racialized risk factors. At the scale of mass screening, the call for every woman to receive a Pap smear became reposed as a project to extend the reach of the Pap smear to raced risky subjects who were newly countable.

Race in the 1980s was thus invoked as a category (by health activists and by health experts) that *explained* differences in cervical cancer rates, requiring efforts to bring racialized constituencies of women into screening. As a historian, I want to ask how debiologized race came to be seen as that which could account for cervical cancer, thereby asking how race as a kind of evidence of risk was produced.[74] In epidemiological work of the 1980s, race was typically not a biological explanation, but instead served as a proxy that could stand in for lifestyles or social-economic conditions, without requiring investigators to ask questions about how those conditions came about in the first place. In this way, racialized difference (rather than, say, sexual history or political economy) was assigned a preeminent epidemiological explanatory function. For example, in the 1980s through today, hundreds of medical studies have focused on differences between blacks and whites concerning cervical cancer, even when the studies' main conclusions are that differences could be just as well explained by income or other factors. Historicizing this efflorescence of research racializing cancer requires seeing race as produced and rearticulated in those moments when it is nominated as a mode of categorization and explanation.

The prevalence of race as the most prominent variable for identifying differences in screening and cervical cancer rates in the United States signals the way, after the fall of official juridical segregation, that race was remade as an object of governmentality—for medicine, the state, patients, raced subjects—and feminists.

Mapping Political Economy, Race, and Risk

It was in response to statistics about racial differences in screening that feminists in the 1980s began to map the problem of cervical cancer and the solution of the Pap smear in terms of a racialized political economy. This was the case in the most widely circulated feminist critique in the 1980s of the Pap smear, *A Feminist Approach to Pap Tests*.[75] In the tradition of feminist self help, much of the booklet concerns detailed attention to the scale of the individual's health encounter in the clinical setting and the practice of information sharing as crafted during a decade of feminist health collectives. At the same time, the booklet's analysis of the Pap smear created a map that highlights a racialized political economy of screening, which in turn is indicative of how feminist self help had changed as a project. Feminists who situated the Pap smear and cervical cancer in political economy did so in multiple ways, featuring different ways of plotting race, risk, capital, and state conjunctures. What counts as the problem of "political economy" has its own politics. According to differently plotted political economies, feminists variously argued that screening should be extended, denounced, or deemed irrelevant.

Continued deaths from cervical cancer, according to the Vancouver-produced booklet, were the preventable failure of mass screening to reach certain constituencies, such as "immigrant women," "native women," and "rural women."[76] At this resolution, what was wrong with the Pap smear revolved around who was and who was not being screened, who was and who was not at more or less risk, and how to adjudicate what made up risk. The implicated subject of screening remained universalized "women" in general, but distinctions were now drawn between different "populations" of women who were missed, variously identified by race, location, class, sexual activity, and citizenship status. The solution to this problem of incomplete screening, then, was to find ways to extend screening, by, for example, educating "immigrants" and "native women" or creating visiting or mobile clinics for rural or underserved areas.

Written in Canada, the booklet is particularly concerned with cervical cancer rates among "native women," who, it argues, were at risk for cervical cancer because they, like the European working class of the nineteenth century, suffered from social and economic, as well as colonial, deprivations: "The lack of jobs, the housing for aboriginal peoples, the destruction of their way of life and the imposition of European culture results in higher death rates from this treatable disease than are found in the general population."[77] Comparing cervical cancer to tuberculosis, the booklet argues that cancer prevention went beyond the Pap smear to include "social and economic patterns." Similarly to how tuberculosis mortality in the nineteenth century was linked to the physical hardships of industrialization—"the social fabric of people's lives was ripped apart"—such that "native people both on reserves and in the cities have the highest rate of tuberculosis." Just as no one needs to have tuberculosis, "no woman needs to be diagnosed with cervical cancer."[78]

Like late nineteenth-century progressive reformers before them, the answer lay in "education and better social and economic conditions."[79] Thus, the booklet presents itself as written for those readers already accountable to mass screening, staking what the historian Peggy Pascoe calls a *relation of rescue*, in which the rescuers granted themselves the moral authority to teach and serve "other" constituencies of women, typically racialized, in ways afforded by already existent hierarchies. In the Vancouver booklet, the relation of rescue is concerned with racialized women outside of screening.[80] What particular map of political economy did the Vancouver booklet thereby produce?

In the clinical encounter, every aspect of the Pap smear had been invested with the exercise of power. Yet, when it came to screening, inequities were the product of how power operated in "society," in "economics," and especially in the past—in "history"—to form the conditions that prevented *other* people from enjoying screening. Unraced feminist self help within the clinic was joined with an analysis of racialized political economy "out there," such that feminist health critics did not often analyze screening program practices as themselves possibly entangled with state racism. In the critical feminist map provided by the Vancouver booklet, the work of race and capital are selectively identified as residing in the structural conditions—colonialism, poor access, poverty—that prevented people from seeking or accessing screening or caused general "stress"— without also focusing on the ways race and capital operated within practices of screening, public health more broadly, and certainly not within feminist clinics.

Thus, feminist interventions into the Pap smear can be seen as participating in late twentieth-century racial governmentality. In other words, they also sought to govern health through a sense of race as a debiologized proxy category of political economy and culture. Quantitatively capturing racialized risk through statistics and epidemiology was a dominant practice of racial governmentality in medical literatures, whereas within critical feminist projects racialized risk was more likely to be marked as the ways inequalities were produced, rather than solved, by exploitative histories of capital and colonialism. Nonetheless, feminists and health experts were joined in their efforts to figure out what process, variable, or inequity was the underlying cause represented by this racial accounting of risk, thereby mutually constituting, even in contestation, the assumption that cervical cancer and screening should be addressed through categories of racial governmentality.

The möbius circuit of accountability in racial governmentality was not only performed by the state, biomedicine, and unraced feminist health projects that selectively analyzed race "out there," but also by health projects explicitly fashioned by and for women of color about cervical cancer. For example, in the early 1990s the National Black Women's Health Project (NBWHP) was part of a collaborative effort to extend screening to low-income black women in Atlanta, particularly women who lived in public housing. With funding from the National Cancer Institute, the project assembled together Atlanta doctors with community health organizers of the NBWHP to design a pilot public health program to overcome the "cultural and logistical barriers" to cervical cancer screening for low-income black women. Here, the medical researchers and black feminist activists were joined by an anthropologist who helped to articulate ethnographic notions of racialized cultural difference.[81] Crafting a coalition between feminist, medical, and ethnographic logics the public health project employed black laywomen as door-to-door recruiters for Pap smears, as well as distributors of "culturally sensitive" educational materials that used black actors and offered an "empowerment" philosophy of health care by and for the black community.[82] At the scale of mass screening, the feedback loop of racial governmentality was redirected by such sophisticated and critical antiracist feminist projects. This project rearranged efforts that saved lives through better spreading the Pap smears, drawing lay black women in as experts in public health knowledge making, and capturing the work of race in history-laden distributions of dispossession. At the same time, the project was possible in the first place because it mobilized the feedback loop of biomedicalized accounts of race as differential risk.

Despite their critical divergence from the tenor of much medical research, feminist self help of the 1980s, from Vancouver to Atlanta (and unlike a decade before), relied prominently on engagements with medical research, even when reformulating them. This was symptomatic of a pervasive change in the 1980s toward more conventional medical structures inside feminist clinics all along the West Coast and across the continent. The rise of HMOs and more stringent state licensing, as well as harassment by a sometimes violent antiabortion movement, all conspired to encourage feminist health centers to more closely conform to the practices of commodified, professionalized biomedicine and its emerging neoliberal formulations. In Canada the Vancouver clinic closed, and the collective that wrote the booklet reorganized as a research and information-sharing center.[83]

As feminist self help had celebrated laywomen as experts in their own lives, it is not surprising that their sense of "expert" shifted in the wake of these changes. The couplets of author/reader, screener/screened, service provider/consumer shifted weight as working in women's health became its own kind of expertise. The feminist self help collapsing of patient and expert can be reread as a constitutive sign of how possible it was for women, particularly white women, to see themselves on both sides of these medical divides as early as 1970. While white men continued to numerically predominate as doctors, the percentage of women physicians in the United States doubled between 1970 and 1980.[84] By the end of the 1980s, one in five physicians were women, though only one in a hundred were black women.[85] With the growing conformity of feminist clinics, it was becoming easier for feminist health providers to recognize themselves as primarily deliverers, rather than targets, of a screening program articulated as a relation of rescue.

Numerous feminist scholars in the 1980s, writing from the recently institutionalized site of academic women's studies, lamented these entanglements between biomedicine, capitalism, and feminism, offering other ways of mapping the political economy of screening. For example, Nancy Worcester and Marianne Whatley, who both held PhDs from the sciences and worked in the women's studies program at the University of Wisconsin, Madison, wrote trenchant critiques of the commodification and remedicalization of women's health under the auspices of new "women's health centers" within established medicine. Such centers advertised attractive feminist features such as women doctors and "pinks and purples" décor, as well as some clinical protocols taken from feminist clinics. Worcester and Whatley argued that dominant biomedicine was "co-opting" the feminist health movement: "Anytime you can develop something to sell to normal, healthy women," they argued, "there is a huge market waiting to be exploited."[86] They particularly singled out screening as "the role of technology in the cooptation of the Women's Health Movement."[87] Pap smears joined mammograms and osteoporosis screening to entice privileged women into bringing their business to for-profit biomedical women's health centers, which further "perpetuates a system which better serves white women."[88] They warned "white middle class women, able to afford the services of the highly visible new centers, will be momentarily silenced by the lure of an attractive range of services."[89] Here, screening itself was mapped as part of a capitalist political economy, in which feminism and white women were becoming enfolded and enticed.

Importantly, this suturing of race, screening, the state, and value not only spread through private gynecology offices and women's health clinics, but also through emergency rooms, public hospitals, and state-sponsored family-planning clinics. Pap smear screening, as much as it was implicated in middle-class health politics, was implicated in the ways public health shifted between valued and devalued citizens, as well as noncitizens. More broadly, public health screening has historically played an important state function of sorting between "life to be protected" and "life to be protected from." For example, in the early twentieth century public health programs at the borders—from Ellis Island to El Paso to Angel Island—had been vital to screening immigrants and defining fit citizenship and valuable laborers on eugenical, medical, and economic grounds.[90] At borders, quarantining, disinfecting, physical examinations, X-rays, and intelligent tests, later joined by HIV tests, were all part of guarding and patrolling who was a viable citizen. Similarly, the biopolitical sorting work of public health extended within the nation. In the state of California, public health practices had systematically demarcated Mexican and Chinese residents as outside and threatening to the nation's white biopolitics, as carriers of infectious disease, as available to harmful labor practices, or as segregable into noxious living quarters.[91] Coercive racist sterilization in public hospitals could act as a kind of biopolitical border control, culling unwanted future lives from citizenship.

In the 1960s, the role of public health in upholding the racial state was crucially transformed through the double trajectories of civil rights and neoliberalism. Health protections were to be extended to racialized citizens at the same time that these constituencies could now be charged in new ways with making themselves available and responsible to biomedicine as a duty of their citizenship.[92] Nonetheless, racialized risks still marked dispossessions layered by earlier decades of racism, manifest in the example of designating Haitians as a risk category for AIDS and the subsequent policing and detention of Haitians at immigration camps.[93] At the same time, in hospitals and clinics, white women as prototypical patients were increasingly joined by the subject-figure of the classed and raced Medicaid patient, who in turn accessed health care through the same logics that marked their difference. In these rearrangements of biomedicine, race, and governmentality, 1965 to 1980 was the period of the greatest improvement in the health status of black Americans overall, at the same time as there emerged new venues for governing racialized forms of citizenship and risk (such as the establishment of the Indian Health Services or efforts to prevent "illegal" immigrants from using health services).[94] By the 1990s, racialized risk could even be exploited as a marketing niche for pharmaceutical commodities.

The Pap smear, which was already attached to family planning and surgical interventions through gynecology, allowed the biopolitical figure of the precancerous patient who rationally makes herself available to biomedicine for the sake of her vitality to be proximate to the necropolitics of population control—that is, the economized logics that designated poor women, and hence often raced women, as the bearers of lives less worth living.[95] The risky figure of the woman missed by screening could be coordinated with the figure of the poor or racialized promiscuous women, and with charges of illicit sexuality, in a moment when the state under Reagan increasingly criminalized reproduction, willfully neglected people with AIDS as outside heteronormative rights, and materially neglected the devastating impact of AIDS on black women even as they were objectified and stigmatized as "carriers" of risk.[96] Under these dispossessing conditions, the spread of family planning nonetheless still offered a means to provide Pap smears to those missed by other venues of health care. The biopolitical work of the state was performed by doctors paid by Medicaid in the name of civil rights to provide Pap smears along with contraception. In these ways, the Pap smear was performed within a complex rearrangement of efforts to preserve life within medical and state structures that selectively withheld health care from purportedly less-deserving others.

The binding of injunctions to health with forms of necropolitics within practices of Pap smear screening is made strikingly manifest within prisons. For example, while incarcerated women in British Columbia had been some of the first recipients of one of the oldest Pap smear screening programs, serving as test subjects to epidemiologically demonstrate the procedure's use-value, by the 1960s pilot Pap smear programs in prisons—from Canada to California to Mexico—were promoted as a means to reach at-risk women. In prisons screening was "relatively easy since no motivating technique is required to bring them to the clinic."[97] Ideally a Pap smear would be incorporated into the initial processing of arriving inmates, as it was in California.[98] Screening at the gateway to incarceration was an injunction to health at the very moment of abjection. It joined the biomedical moment of care in the name of "women's health" to a mandated opportunity for subjection and humiliation. Cervical cancer screening could offer a conduit for sexualizing incarceration.

The confluence between biomedicine, women's health, and the racial state—between privatized and gendered calls to consume health and new racial formations governed by cost, benefit, and risk—helps to reveal as constitutive a seeming paradox about Pap smear screening and the distribution of cervical cancer mortality. Though black women received screening as often or more so than white women did, they died, and still die, of cancer more often. Where and under what conditions a Pap smear was taken had an effect on the bipolitical work it did.

Within screening, "race" became a form of accounting for distributions of life and death within stratified biomedicine, differentially marked and mapped within a proliferation of feminisms. Beverly Smith, a member of the Combahee River Collective and university lecturer, and Angela Davis, a notable black feminist and communist, now a professor, both offered a feminist political economy of race and health in the 1980s. Asked if the problem of black women's health was about health care, Smith answered,

> A lot of people don't understand that the availability of medical care is not the primary thing that impacts health status. Economic and social forces such as good nutrition, good housing, a clean water supply, adequate clothing and sanitation influence health care the most. Having adequate access to those things is going to go much farther to enhance your health status than lots of medical care. . . . If the people you want to treat are drinking contaminated water, living exposed to the elements or not getting proper nutrition, then the health center really isn't going to help them much.[99]

Smith preferred to articulate the causes of black women's health disparities in terms of lack of "freedom and safety." Neither single health issues, nor even health on its own, Beverly Smith argued, galvanized the politics of most black women: "It is a luxury to be able to focus on a single political issue because so often one's life is about surviving a host of different oppressions and then dealing with all of the problems and struggles at once"[100] Audre Lorde, a feminist theorist, poet, and member of the

Combahee River Collective, wrote about her cancer experience in the acclaimed *Cancer Journals* as well as in short form in many other venues, theorizing her own "battle" with breast cancer similarly: "Battling racism and battling heterosexism and battling apartheid share the same urgency inside me as battling cancer."[101] Angela Davis's theorization of the politics of black women's health in the 1980s situated it on racialized biomedical, nation-state, violent, and economic coordinates. Not only was medicine actively racist, denying treatment to black Americans, but so was the state under the Reagan administration, which increased military spending along with the privatization of state services that "prioritized the profit seeking interests of monopoly corporations" over citizens' lives. Davis emphasized that over a quarter of blacks receiving health care in 1983 did so in emergency rooms.[102] For Davis, access to medicine took on dramatically different valence than in the Vancouver booklet: "We must learn consistently to place our battle for universally accessible health care in its larger social and political context."[103] On the map of "simultaneous oppressions" of the 1980s, cervical cancer screening programs were not highlighted in their specificity and the Pap smear was neither the right or the wrong tool, but a largely irrelevant one.[104]

In sum, the scale of screening created disquieting proximities between feminist health practices and racial governmentality. State, medical, and feminist calls for "every woman" to have "access" to a Pap smear that most, but not all, women could afford to enjoy were connected to the unraced flexible inclusiveness of feminist self help's clinical strategies, which were in turn folded into the political economic analysis of racialized risk, which in turn expressed racial governmentalities, which in turn joined the promotion of life with the governing of racialized dispossession. As I've tried to show, how feminists mapped the problem of cervical cancer amidst these manifold relations itself involved an ontological politics of "political economy" (rather than of organs and cells) as the nature of the problem. "Political economy," was mapped by discrepant feminisms who varyingly materialized some relations and not others, for which the Pap smear could both be crucial and beside the point. Political economy, as a problem-space, could be mapped by feminists as raced and classed "social and economic patterns" "out there" requiring enrolling risky women into the "in here" of screening. Or, political economy could be invoked by proxy, but not explained, in the epidemiological use of the category of race within a reconfigured biomedicine. Or, political economy could alter the resolution of analysis, mapping cancer into a topology of simultaneous, interlocking oppressions, for which the Pap smear as a feminist project receded from view.

Becoming the Wrong Tool, Reproductive Health, and the Tangle of Transnational Feminism

Epidemiologists who studied cervical cancer had long made connections with sexual activity, leading to speculations of a yet unidentified causal infectious agent. Just such an agent was conclusively identified in the early 1990s: the human papilloma virus (HPV). It is a large family of viruses that infect the skin and are passed by skin-to-skin contact, with different strains variously causing problems from common warts to genital warts to cervical cancer to no symptoms at all. A subset of HPV types are estimated to be the cause of 99.7 percent of cervical cancers worldwide, leading experts to wonder if there is such a thing as cervical cancer without HPV.[105] At the same time, most HPV infections do not necessarily lead to cancer. The ontological politics of cervical cancer had thus rearranged again. Recognized as caused by a sexually transmitted infection, it was no longer centrally a problem of errant cells but of risk caused by a virus.

Much of the research to verify the cause of cervical cancer and strains of HPV has been conducted in "developing world" locations, such as Zimbabwe, South Africa, Mexico, and Costa Rica, among women who had never had Pap smears.[106] Such research sites were not hard to find. While cervical cancer had declined in the United States, Canada, and other countries of the so-called global North with elaborate health care systems, cervical cancer had also increased in previously colonized countries in the global South that lacked such health care infrastructures. The success of the Pap smear in the United States and elsewhere had depended on both an industrialized laboratory network and a means to treat women with positive readings. It depended on equipment—from swabs to fixatives to microscope, as well as lasers and operating theaters for treatment.[107] It depended on infrastructures of clinics, trained specialists, and transportation, as well as clients able and available to present themselves for screening. Moreover, it depended on funds from individuals, states, NGOs, or insurance to pay for all these aspects. This constellation of conditions was distributed such that most of the world's women were excluded from the regime for apprehending cervical cells that had become commonplace in the United States and Canada. In many places, the equipment to take, the infrastructure to read, and the ability to treat cannot be or were not assembled, making it impossible for women to screen for early cancer. On the profoundly uneven transnational terrain of health infrastructure, cervical cancers became one of the four "cancers of underdevelopment"— cancers that were rare or decreasing in previous colonial centers but common or increasing in postcolonial "developing" countries.[108] Thus, while cancer is often presented as a disease that strikes rich and poor alike, its fatal results are far from egalitarian.

The Pap smear, on this transnational scale, had not been the right tool for the job. Cervical cancer remained one of the most common cancers for women in the world, according to the World Health Organization, with 80 percent of all cervical cancer deaths in "developing" countries, making cervical cancer the biggest cancer killer of women in the poorest countries of the world.[109] In such international health measures, cervical cancer death rates were measured by nation, marking a geopolitics in which the lack of health infrastructure allowed cervical cancer to kill to a greater extent. Cervical cancer, moreover, was a very different kind of problem than breast or lung cancer; it correlated with a sexually transmitted infection and was more easily treatable. Its ontological politics had been rearranged yet again. And as such, by the 1990s, in the shadow of transnational HIV and family-planning projects, HPV was not just infectious but a feature of "underdevelopment," and hence amenable to a new host of fixes, from safe sex with a condom, to the promise of vaccines, to empowerment.

At the scope of the transnational, I will track how feminists mapped interventions into "cervical cancer as infectious disease" by fashioning new forms of transnational relations of rescue and yet another twist to ontological politics. The Pap smear once again was an important juncture, helping to rematerialize cervical cancer as solvable within a "reproductive health" global policy that required a rearrangement of the transnational family-planning industry. Important to this story was the New York–based International Women's Health Coalition (IWHC), founded in 1984 with funding from the Population Crisis Committee. The IWHC was a feminist, nonprofit organization that by the early 1990s was one of the largest, most influential feminist NGOs. It was led by Joan Dunlop, who had been an advisor to John Rockefeller III on his population control policy in the 1970s, and Adrienne Germain, who was the first woman to hold the position of country representative (for Bangladesh) within the Ford Foundation. Neither came from grassroots feminist health activism in national or local spheres; instead IWHC took as its home territory international population policy, seeking to transform the protocols of population control projects whose funding more often than not originated in the United States. Protocol feminism had gone transnational.

One of the IWHC's earliest projects involved reformulating the ontological politics of cervical cancer. Working with Judith Wasserheit, a Center for Disease Control medical researcher, IWHC developed the concept of "reproductive tract infections" (RTIs) in the late 1980s.[110] The term *reproductive tract infection* held together a variety of infections from not only sexual intercourse "but also from the use of unclean menstrual cloths; insertion of leaves and other materials in the vagina to increase a male partner's pleasure, prevent pregnancy, or induce abortion; unsafe childbirth or abortion techniques; and other harmful practices such as female circumcision."[111] The IWHC's goal of grouping cervical cancer with herpes, candidiasis, and sepsis from medical procedures was to create a single, sweeping rubric that constituted a new problem. A general category of reproductive tract infections might stimulate a reallocation and reorganization of health services (which were then overwhelmingly oriented toward fertility control) for women in the "third world." With cervical cancer newly identified as linked to a sexually transmitted virus, and with the history of the Pap smear as a cornerstone of preventative gynecology in the United States, it made strategic sense to move cervical cancer into the category of RTI in order to prod the entrenched transnational infrastructure of family-planning services to also offer basic health care such as Pap smears.

The IWHC's efforts to craft and promote the concept of RTIs signaled an emergent feminist strategy that saw collaborations between geopolitically distant NGOs in order to target United Nations conferences and policies as points of intervention. The existence of the International Women's Health Coalition was a symptom of a historical change in the organization of feminist health activities and in the form of protocol feminism both in the United States and transnationally. The scholar Sabine Lang has called this the "NGO-ization of feminism," in which the dominant mode of feminist organizing became the acronym-filled universe of the NGO.[112] Growing NGO-ization was accompanied by a proliferation of the spaces and places in which feminisms were articulated, and an increased presence of women from the "global South" in them.[113] Such nongovernmental organizations formed national and transnational networks, often organized by a new kind of feminist expert whose terrain of politics was the policies, guidelines, and protocols used in service provision and disseminated by supranational organizations established in mid-century: the United Nations, the World Health Organization, the International Monetary Fund, and the World Bank. The NGO-ization of feminism, in turn, resonated with the neoliberal rearrangement of health services into the private sector and out of the state. At the same time, this feminist formation built on the local legacies of various national women's health movements, expanding the scope of protocol feminism from local feminist projects to the policies that distributed and managed health care and development.

The crafting of strategic transnational coalitions into international "women's caucuses," "women's coalitions," and "networks" supporting a common platform became an important tactic.[114] American feminist organizations clustered on the East Coast—such as the IWHC—played a particular kind of role in these efforts in the early 1990s, including fundraising, funding other NGOs, and serving as secretariats for preparatory meetings, where they conducted the laborious work of drafting, then circulating documents in a way that incorporated feedback and contributions from less-resourced NGOs working in far-ranging sites and under diverse situations.[115] In these processes, Southern and grassroots NGOs navigated the uneven expressions of NGO-ization and the more dominant role of American, professional NGOs in it. American feminists, in turn, unevenly met the obligation to distribute leadership and resources in equitable ways that did not just maintain Northern NGOs in the position of gatekeepers. In this contradictory terrain, the IWHC straddled hegemonic and counter-hegemonic spaces in their professional life. Thus, in this historical shift to the NGO-ization of feminism and health care, the IWHC stood in a particular kind of place.

What the place of the "transnational" is, is not self-evident, and the IWHC posited a particular map of its extension. The world was divided into the binary categories "North" and "South," without finer resolutions of stratigraphy inside of regions or nations. Moreover, North and South mapped onto the categories of "developed" and "undeveloped," thereby echoing cartographies of the world as riven according to economic stages of progress, which in turn rested on and reframed divisions of the world into the colonial logics of modern and traditional. Moreover, the broad sweep of this cartography allowed gross generalizations across and within nations that bound diverse women in the "South" into common narratives and explanatory frameworks. Although feminists did not invent this map, they nonetheless invoked it, even while rearranging it.

This map of "developed" and "undeveloped" was already the stage of supranational and national endeavors to curb population growth through the dispersion of contraception and sterilization — family-planning projects that had become a central feature of postcolonial relations of rule between decolonizing countries and emergent new imperialisms that encouraged the spread of global capitalism.[116] Vast amounts of United States funds had been spent since the 1960s on projects to limit the reproduction of poor women, a flow of resources that feminist organizations strategically appropriated. The International Women's Health Coalition, with its own multimillion dollar budget collected from Mellon, Ford, Hewlett, and Macarthur foundations, USAID, and wealthy private donors, did just that.

The IWHC initially tended to spend its funds on, first, organizing transnational conferences around articulating a common feminist "reproductive health" policy vision and, second, on funding and providing assistance to local feminist women's health NGOs — what it called its "colleagues." As Adrienne Germain described, "We found and invested in like-minded individuals and organizations across Africa, Asian, Eastern Europe and Latin America, empowering them, just as they would empower the girls and women in their community."[117] Empowerment, as the scholar Barbara Cruikshank has argued, is a specific political discourse articulated since the 1960s that rests on practices of regulating participation, power, and political subjectivity. More specifically, "empowerment" is a strategy for regulating political subjects by seeking to foster the capacities of the "powerless" and maximize their participation towards a particular end.[118] By the 1990s, feminist NGOs appropriated the term *empowerment* to describe efforts (in a direct lineage to feminist self help) that sought to give their clients capacities for governing their own reproduction, health, and work. At the same time, empowerment was fashioned through hierarchical relations of rescue.

Navigation of this terrain by the IWHC was as self-identified "Northern" feminists working within what was becoming one of the most endowed feminist transnational NGOs. To develop and promote the concept of "reproductive tract infections," IWHC members organized a Rockefeller-funded conference at the prestigious Bellagio Study and Conference Center set in the beautiful hills of northern Italy, inviting eminent researchers in the field of sexually transmitted disease, journalists, population control policy makers, and representatives of international donor agencies.[119] They also organized a panel at the 1989 American Public Health Association Meeting. Here, the audience for the notion of RTIs was one of biomedical experts. The concept of RTIs articulated in both these venues was an initial testing ground for later developing the policy concept of "reproductive health." Leadership at the IWHC argued that the "reproductive health approach, with women at its center, could considerably strengthen the achievements of existing family-planning and health programs, while helping women to attain health, dignity and basic rights."[120] It involved joining programs already extant under the rubrics of family planning, child survival, safe motherhood, women in development, and health into a single integrated "comprehensive services" that "empowered women to manage their health and sexuality" under the banner of "reproductive health."[121] At this point, attending to RTIs was a concept of protocol feminism, meant to rearrange and refocus the organization of health services.

The IWHC then planned another conference in March of 1992 that radically changed and expanded the concept of RTIs. This time they collaborated with the Woman and Development Unit (WAND) of the University of West Indies, then headed by Peggy Antrobus, meeting in Barbados. Antrobus was a prominent feminist expert in the Caribbean, serving not only as a professor, but as director of the Development Alternatives with Women for a New Era (DAWN), a transnational network of women in the South organized around issues of development, and as the previous head of the Women's Bureau of Jamaica. Antrobus's work, like that of many feminist experts, saw her moving across these contradictory spaces as administrator, researcher, activist, and critic. Unlike at Bellagio, at the IWHC/WAND conference participants were all women: forty-four activists, physicians, NGO representatives, scholars, and journalists, mostly from the Caribbean, Latin American, Africa, and Asia, as well as a contingent from the United States, who drafted a "call to action for new alliances between women and men" concerning RTIs.[122] In this document, the deadly fact of cervical cancer had considerable prominence. The document argued that RTIs, including cervical cancer, had been failingly approached as "diseases to be mapped by epidemiologists, prevented through public education campaigns, and cured by health professionals." Instead, they advocated understanding RTIs in terms of the "power imbalances" in both the private and public spheres, with particular emphasis on the "pervasive imbalance of power between women and men." Moreover, these "power imbalances" were also found in the "present global political and economic context," including the shift by "northern governments, multilateral institutions and Southern governments" from human development to "privatization and economic growth."[123]

Though cervical cancer was linked to a sexually transmitted virus, the concept of RTI foregrounded that risk was not merely from sexual behavior, but was an outcome of the uneven power relations that governed sex, as well as health care. The conference report mapped the "imbalances of power between men and women in virtually every society" as a primary cause of RTIs, preventing women from protecting themselves or seeking treatment.[124] In particular, participants named a "culture of silence"— which became a major theme in the later IWHC campaign—that had the dangerous result of stopping women from critiquing their circumstances and seeking health care. They concluded that conventional interventions into RTIs, "for example providing women with income, information, health, education, and services," would "be effective only if the imbalances in gender power relations are directly addressed."[125] Finally, building on critiques of the value of condoms in preventing the transmission of HIV for women, the IWHC-WAND document called for the development of new technologies "such as vaginal microbicides that can prevent disease transmission without preventing wanted pregnancies and that a woman herself can use without her partner's knowledge or consent."[126] Directing their call at donor agencies as well as policy makers and feminist activists, they urged that "health, including sexual and reproductive health, be perceived and treated as a basic human right."[127]

The IWHC's own later descriptions of RTIs and "reproductive health" in the early 1990s tended to emphasize the provision and quality of health services in family-planning clinics and the need for reproductive rights, and did not continue to highlight the critique of political economy. The WAND, on the other hand, working with the Latin American and Caribbean Women's Health Network, developed its own campaign, "Demystifying and Fighting Cervical Cancer."[128] The campaign was organized by the activist Andaiye, who had a quite different history of political organizing from WAND, though they both shared a critique of structural adjustment. As a critical Marxist Guyanese grassroots feminist, she was a founding member, along with Walter Rodney, of the Working People's Alliance political party; and a cofounder in 1986 of Red Thread, an organization that forged a multiracial politics around valuing the unwaged labor of poor women, and would later become a significant force in the Global Strike for Women.[129] Andaiye had herself had cancer, and was also a friend of Audre Lorde, the author of the renowned *Cancer Journals*. Along with her collaborator Selma James, Andaiye's politics might be called "left of Marx," in that it sought to radicalize anticolonial Left politics with questions of sex and race.[130]

With Andaiye as coordinator, the campaign opened with a political economic analysis of the uneven distribution of Pap smear screening, speculating that "Barbados could have the highest rate of cervical cancer in the world."[131] The question of "risk" was dramatically reframed, not in terms of individuals, but as "the imbalance of power between rich and poor, North and South, white and nonwhite, men and women."[132] Moreover, they mapped this political economic distribution of risk between North and South, as "also aris[ing] from the priorities of Southern Governments—or, rather, from the acceptance by Southern governments that the priorities for our countries and people are properly set by the North. Hence their adoption of International Monetary Fund and World Bank structural adjustment policies, which could contribute to increases in cancer and cancer mortality in the South."[133] The politics of risk was implicitly problematized as a calculus of already devalued human life that resulted from fostering the "productive" sectors of the nation and not human life itself. This political economy of risk, the WAND pamphlet emphasizes, was interested in "garnering foreign exchange at virtually all costs, leading to 'economic' decisions that increase pollution and the exposure of our populations to pollution . . . This goes hand-in-hand with the continued absence of legislation to protect workers and communities."[134] On this cartography of exploitative transnational economics that posited Barbados as a site for cheap labor, the WAND pamphlet remaps the question of empowerment in political economy: "It is important for us to recognize that an individual's choices are made within a context shaped by larger forces. Human behavior is a reflection (although not a mirror image) of material life, and our material life is shaped by the power relations in which we live."[135] Feminist health campaigns thus called for "self-determinism" at both the level of the individual, the community, and the nation. In its campaign against cervical cancer, Andaiye and WAND created its own reassembly, inserting analyses and images of the Pap smear copied from the Vancouver booklet, together with RTIs, and a map of neocolonial political economy, calling upon the reader as both a potential sufferer of cervical cancer, and as a potential organizer of her own local campaign.[136]

While collaborating with NGOs such as the WAND, the IWHC also straddled relations with northern policy and research insiders who foreclosed questions of political economy. At this contradictory conjuncture, the IWHC posited the problem of cervical cancer as involving more than dividing the world into North and South; it presented the problem as solvable at the overlapping concern of revising gender roles — calling for gender to be a new site of intervention and governmentality — and a reallocation of health policy, health care, and family planning. Unlike the WAND literature, the IWHC publications did not highlight the political economy that had made the Pap smear the wrong tool in the first place. Rather, the IWHC's formulation of RTIs foreshadowed some of the ways they later constrained "reproductive health" as a new international policy protocol.

The IWHC is best known for its work behind the successful approval of "reproductive health" as an official policy goal at the UN International Conference on Population and Development (ICPD) held in Cairo in 1994.[137] The conference platform replaced neo-Malthusian language (regulation of birthrates in the name of reducing poverty and improving economies) with the goal of "reproductive health" in the approved text. In many ways, the IWHC's previous work on cervical cancer and "Reproductive Tract Infections" had been a step toward the group's strategies at Cairo. For example, IWHC had organized and helped fund a small transnational conference of prominent women's health advocates from both the North and the South that created a draft of what would become the women's caucus platform for the ICPD, a draft that was then circulated and rewritten across even more NGOs.[138] Over twenty-two thousand signatures eventually approved the final statement. The IWHC, then, funded by the Ford Foundation, organized another conference, with 215 representatives from 79 countries in Rio de Janeiro, where the strategy of coming to the ICPD with a united women's caucus was secured.[139] With one foot in the world of feminist NGOs and another in the world of policy makers and East Coast elites, Adrienne Germain was appointed as an official delegate of the United States to the conference, where she held a "war room" that helped to choreograph the translation of the feminist platform into the official approved UN policy.

The dramatic change in policy achieved at Cairo — the embracing of "reproductive health," with its emphasis on the quality of care, equal rights for women, and the expansion of family planning beyond demographic targets to include issues of sexual and reproductive health—has provided guidelines for redirecting vast currents of resources toward a new formulation of the "problem" of women's health. In this formulation, development could not be achieved without women, and hence without feminism, and thus family-planning services had to be delivered as part of the larger category of reproductive health, which in turn necessitated "empowering" women. What was missing, critics pointed out, was any alteration in the neoliberal macroeconomic logics by which "development" was proceeding.[140] Development, in its neoliberal form, and reproductive health became hitched together as a conjoined transnational relation of rescue intent on altering "gender" and "investing in women."[141] A new formulation of transnational gendered governmentality was emerging, and the question of the Pap smear was enveloped within. By the end of the century, for feminists, nation-states, the UN, and the World Bank, "gender" and "reproductive health" had become technocratic objects of intervention, investment, and governance.

What kind of cartography was crafted by the IWHC, and not necessarily shared by its collaborators and allies, in the politicization of cervical cancer into reproductive tract infections? Power was portrayed as something imbalanced, between men and women, and between North and South. The IWHC saw itself as situated in and intervening in this imbalance by calling for policies that redistributed resources in the form of "empowerment," by forming coalitions with women from the South, and by calling for women to be actors in policy-making venues, in programs, and in all aspects of decision making, including at the individual level.[142] The IWHC was highly aware of its position within the "North," yet nonetheless fashioned itself in a chain of "empowerment" with itself at the apex of a disbursement of financial flows and policy recommendations. This chain of empowerment might also be read in reverse gradient, as making possible a web of local appropriations of resources facilitated by the rubrics of reproductive health and empowerment.

The yearly galas the IWHC throws as fundraising events for New York socialites are revealing of some of the contradictions in the politics of "empowerment." Bedecked in jewels and black pantsuits, elite donors—almost all white—were entertained in a chandeliered banquet room by white tuxedoed waiters, while awards were given to Hillary Clinton (2005), and Kofi Annan (2004); Hollywood actors offered speeches; and Paparazzi snapped photos. At the 2005 gala, the walls surrounding donors were decorated with large illuminated photos of the unnamed brown women and girls to be empowered (see figure 3.4). The convolutions that the IWHC inhabits were further made visible, when in 2007 two female members of the De Beers Group, the South African Diamond trade titans, were invited to give prominent speeches as supporters and donors. At such events, the IWHC regularly raises over a million dollars, and thus they are important to the overall budget. As one of the board members explained at the 2006 gala, "We are the trusted partner of both, the powerful and the power-less."[143] Or, as one of the De Beers spokeswomen explained in her 2007 speech (republished on the IWHC website):

> As a woman I am very proud that De Beers is supporting IWHC. Almost as proud as I am as an African to be working for De Beers. I am proud because my father said, "I have seen what diamonds can mean for Africa," and because De Beers as a company believes and practices the principle that Africa needs a hand up, not a hand out. I have seen more than a glimpse of Africa developing to its full potential. I believe that being here, you, the individuals and companies represented in this room, are demonstrating that you are not satisfied with a tiny glimpse. You, like us at De Beers, are finding new ways to grow potential by supporting the health and rights of young girls worldwide. . . . At De Beers we call this "living up to diamonds."[144]

Here accumulation and dispossession joined together as diamond mines are mobilized toward the empowerment of women in reproductive health. Reproductive health as a capital formation twists yet again. "Living up to Diamonds" captures the emergence of a contemporary feminist biopoli-tics that "lives up" to the logics of global capital.

With the IWHC's position straddling the worlds of elites and NGOs — a metaphor of two feet in different worlds that the IWHC regularly uses — its formulation of reproductive health tended to bracket questions of macroeconomics and instead relied quite squarely on a strategy of empowerment. This was the critique many feminists have made of the Platform of Action won at Cairo, that it allowed a feminist face to be put over what essentially remained the same neoliberal economic policies.[145] Looking back at the rise of the feminist NGOs with "gender experts" that in-

3.4. Photograph of the décor at a 2005 IWHC fundraising gala held in New York, conveying the "relation of rescue" triangulated among elite donors, the IWHC, and the representations of women needing "empowerment." The galas are regularly covered by Patrick Columbia's New York Social Diary website (NewYorkSocialDiary.com). Photography by Jeff Hirsch.

terpreted, advised, and often disguised the neoliberal economic policy pursued by the International Monetary Fund (IMF) and the World Bank (WB), Andaiye critically observed, "And those of us who were not academics were raised up to be consultants. . . . But the gender expert is the same as the race expert is the same as the class expert is the same as a pimp! . . . there is no way that putting women in an IMF/WB document can make it into anything that is friendly to working people."[146]

Like most development projects—and unlike the ways reproductive health was formulated by antiracist feminists within the United States and the Caribbean—the IWHC analyses almost never discussed "race."[147] While identifying itself as from the North, the IWHC did not identify itself as working from a location in the heart of what was a new kind of imperialism, nor as noninnocent participating in the economizing logics of global capitalism. The metaphor of straddling two worlds literally stepped over the contradictions within the NGO-ization of feminism. In fact, it was not until the election of George W. Bush in the new millennium that the IWHC began to explicitly and publicly denounce American "imperialism" in the form of the defunding of reproductive health projects, and the rise of abstinence programs and other Bush administration policies that the IWHC referred to as a "Bush's Other War."[148] Tellingly, the IWHC had, during the Clinton administration, enrolled the president as a supporter of the language of reproductive health and rights and fostered a close connection with Hillary Clinton. The lack of coalition between the IWHC and, for example, the U.S. Women of Color Delegation to the ICPD, or even between the IWHC and grassroots feminist health projects in the United States, reinforced the IWHC's articulation of an unraced feminism located in the so-called North and created a cartography concerned with a binary, transnational world unopposed to macroeconomic development logics, holding at a distance the uneven biopolitical conjunctures within the United States or within the Caribbean.

The goals outlined in the Platform of Action attained in Cairo, all actors would agree, were not met in the decade that came after, yet "reproductive health" has succeeded in becoming a frame of governmentality. Within the IWHC, the concept of the RTI was generally eclipsed by the concept of "reproductive health." The concept of RTIs, however, was widely incorporated into the reconceptualizations of reproductive health crafted at such institutions as the World Health Organization and the Population Council. The "problem" of reproductive tract infections took on a biomedicalized life of its own as an object of inquiry, governance, and rescue, where Pap smears, HIV programs, and family planning became conjoined. Pilot programs have suggested less resource-intensive ways of screening for cervical cancer, such as simple visual examination using vinegar, while new risk assessments found that just one Pap smear in a woman's life, between the ages of thirty-five and forty-five, was enough to dramatically reduce deaths.[149] In the years around the new millennium, an abundance of manuals and managerial programs offered instruction in how to have quality of care and foster reproductive rights in underresourced family-planning clinics, thereby reinserting back into transnational protocol feminism the terrain of the clinic and practices of feminist self help. By 2005, the newest WHO report announced an intensification of cervical cancer deaths in the "developing world" from 80 percent to 95 percent.[150]

Vanishing Pap Smears

The American Association of Cytopathology commissioned a 2007 report on the fate of the cytopathology profession in the face of declining markets for Pap smears.[151] For this professional evaluation, Pap smears were now clearly a commodity and cytopathology a business. Cervical cancer risk was a market. A new technological fix was for sale, rematerializing this market of "every woman" in terms of risk for HPV infection, rather than as cancerous cells to be monitored. The Merck vaccine for HPV — Gardasil — was approved by the FDA in 2006.[152] Merck aggressively lobbied state legislatures to make HPV vaccines mandatory for all girls between nine and twelve. Only in Texas, with direct donations to the governor, did Merck briefly succeed.[153] Mobilizing uneven biopolitical topologies, Garadsil became mandatory in 2008 for young female immigrants to the United States between the ages of eleven and twenty-six, even though it was only recommended by the CDC for U.S. citizens.[154] Merck, as well as other companies associated with the HPV vaccine, also enrolled female congressional representatives through funding "unrestricted educational grants" to their advocacy group, Women in Government. Not just in the United States, Merck lobbied for public legislation in many other national "markets." Nonetheless, the cost for the vaccine is increasingly covered by Medicaid, insurance, and other national health care systems. A new biopolitical formation was entangling with feminist health politics.

With its biopolitical advertising slogan "I want to be one less," and commercials with girls on skateboards or jumping rope chanting "I could be one less statistic," Merck sought to capture the market made possible by the Pap smear but push it into girlhood. Gardasil materials cleverly appropriated activist discourse with catchphrases like "Do something today," "Make an impact," and "Tell a friend" to virally encourage through You-Tube the affective labor of young women as Gardasil promoters. In the *Calcutta Telegraph*, the vaccine was heralded with the headline "Killer at Large," marking cervical cancer as a public health threat in need of vaccination at the same time that Merck outsourced the clinical trials for its vaccine to the Bangalore Clinical Research Organization.[155] In this way, Indian women were called upon as both the experimental biocapital from which the vaccine was developed and the market for its global dissemination. Lifesaving takes on involuted biopolitical forms.

Through the vaccine, designed for pre-HPV exposed girls, risk is temporally moved backward in the life cycle to girlhood, a period of life premised as presexual. Here, with the vaccine, the problem of cervical cancer, feminism, the Pap smear, and risk is remade again as yet another formation of capital tied to an earlier life cycle moment. At these rearranged conjunctures, risk incites business to produce pharmaceutical commodities that in turn are distributed and embraced as life-giving acts for children who embody speculative sexualities of the future.

Feminists have both critiqued and applauded the vaccine as a solution to the problem of cervical cancer. The vaccine has been greeted with suspicion by the Canadian Women's Health Network, while both the International Women's Health Coalition and the Barbados Family Planning Association herald the vaccine with excitement. In proliferating biopolitical conjunctures, Christian groups, antivaccination groups, and corporate watch NGOs jointly denounced Merck. One Christian abstinence group in the United States cleverly reappropriated the Merck advertising slogan in its own YouTube commercial, objecting to the vaccine's implicit sexualization of girls: "I want to be one less object being used."[156] The contradictory entanglements continue.

By tracking feminist interventions of the Pap smear as a question of differently mapped, yet layered, scales, I want to suggest that the ontology of the "problem" of cervical cancer, and health more broadly, are wrought out of historically specific technoscientific, racialized, and spatialized solutions. Rather than a single feminism, or kinds of feminism, I've tried to show a proliferation of sometimes antagonistic feminisms that variously situated themselves relative to technoscience. Moreover, feminisms of many forms have become constitutive elements in these biopolitical—and necropolitical—topologies that have layered over the twentieth century. The scales of the clinical encounter, of the screening program, and of the transnational each traced and placed feminists and the problem of cervical cancer in dramatically discrepant and complexly contradictory ways.

What I have set out to do as a historian is to provide yet another cartography—a map of entanglements. This chapter's map situates various American, Canadian, and Caribbean feminist projects, more in their aspirational forms than in their messy realities, within the histories of racial governmentality and the rise of neoliberal governance. Moreover, these histories have animated not just racist technoscience, not just stratified technoscience, not just normal technoscience, but also feminist and anti-racist politics.

All the feminist projects I have outlined here have accumulated and echoed in the twenty-first century. Feminist NGOs around the world, as well as grassroots projects in the United States, fashioned versions of self help practice. Feminists in the United States and elsewhere still called for better screening. Feminists in the United States and elsewhere still sought ways to bundle cervical cancer with reproductive health—and reproductive health with development. Feminists still tracked biomedicine as a political economy, plotting and navigating the exercise of capitalism.

At stake in all these reassemblies of the Pap smear and cervical cancer is ontological politics. How are "problems" brought into perception to become actionable, to become objects and relations to be named, governed, acted on, and intervened in? Problems are in part fashioned out of the very solutions that presuppose them. But they are not produced out of nothing—they are remade, refolded, rearticulated, rematerialized by assemblages operating at different scales, on discrepant terrains, with shifting practices, that extend and connect in ways marked and unmarked far beyond the moment of swab contacting flesh. The work of bringing a problem" into being is also inevitably, as is all work, the exercise of power on uneven conditions. And how this exercise is imagined, performed, marked, and unmarked is also part of the ontological politics of bringing a problem into its solution. What kinds of cartographies could be drawn, what kinds of tools might be forged, to map the problem-space of cervical cancer, which is also the problem of yawning deadly disparity that does not just explain but needs to be explained? Cervical cancer is at once a mass of fatally proliferating cells, a preventable flaw in screening, an infectious disease, a reproductive tract infection, a sign of gender inequality, a racialized risk, a symptom of underdevelopment, a material consequence of pitiless global capitalism, and a market waiting to be vaccinated. What singular feminism, what politics, can name this problem?

CHAPTER 4

Traveling Technology and a

Device for Not Performing Abortions

A large syringe and a flexible straw—these are the basic components of a device that can suction the contents from a human uterus (see figure 4.1). In the 1970s, this device traveled between Shanghai hospitals, an illegal abortionist in Santa Monica, mostly white American radical feminists, population control policy makers in Washington, D.C., physicians in the International Planned Parenthood Federation, and "lady village workers" in Bangladesh. Its itinerary carved out an emerging biopolitics carried out by the United States in response to the Cold War—a governmentalized investment in intercollating life and economy in decolonizing locales. It was distributed by the tens of thousands to Pakistan, Indonesia, Korea, Singapore, Vietnam, Thailand, and especially to Bangladesh.

The device was simple. It consisted of a 50 cc plastic syringe and flexible polyethylene straw-like cannula that could, in a matter of minutes, suction out the contents of the uterus safely, and without the need for a sterile operating theater. The technology could perform a biopsy, quickly remove menstrual matter, or provide "postabortion" care (the euphemism currently used for finishing an incomplete abortion). It could also empty the uterus of a pregnancy. Moreover, in this final use, the technology was not only portable, easy to learn, and relatively safe, but it was materially constrained to the first weeks of pregnancy which, in the 1970s, was a window *before* pregnancy tests were widely available, and thus a window in which it was often impossible to objectively prove pregnancy.

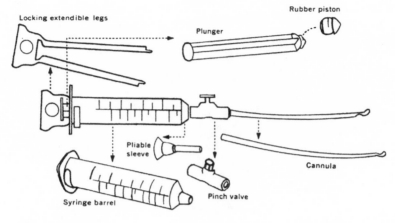

Locking extendible legs

Plunger

Rubber piston

Pliable
sleeve

Cannula

Syringe barrel

Pinch valve

SOURCE: The Pathfinder Fund

4.1. Image of Menstrual Regulation device components: a syringe with pinch valve and cannula, from the Pathfinder Fund. The Pathfinder Fund, supported by USAID grants, developed a menstrual regulation protocol as well as participated in training programs in the 1970s. From Leonard Laufe, "The Menstrual Regulation Procedure" (1977).

Within the United States and its Cold War extensions of foreign aid, this technology went by an assortment of names: uterine aspirator, endometrial aspirator, manual vacuum aspirator, menstrual induction, lunchtime abortion, and miniabortion, but the two names I will focus on were Menstrual Extraction and Menstrual Regulation.

Menstrual Extraction (ME) was the term coined by the Los Angeles–area feminist health activists to describe "the technique whereby a woman can control her menstrual period" (see figure 4.2). In contrast, as used inside the emerging Cold War cartography of family planning and foreign aid, it was called Menstrual Regulation (MR). Menstrual Regulation named a mass-produced commodity that "evacuated the uterine contents from a woman who is at risk of being pregnant, before she can be declared 'obviously pregnant' by clinical examination and other diagnostic measures."[1] While feminist practitioners vociferously asserted that ME was entirely distinct from family-planning MR, what the two instantiations of this device shared was an assertion about what they both were not. They were not methods of abortion. The names that have accrued around this device caused confusion: Were ME and MR the same thing? Were these

Lorraine Rothman, Developer of Menstrual Extraction Kit Del-Em.

4.2. Mimeographed photograph of Rothman with her Menstrual Extraction kit. From FWHC, *Abortion in a Clinical Setting* (1974).

names simply euphemisms for abortion? What counts as an abortion? What makes an abortion feminist?

This chapter takes the feminist claim of difference between ME and MR seriously, but also engages it critically. Tracking ME through a historical lens draws out how the biopolitical project of feminist self help was uneasily entangled within—not outside of—the larger enactment of a Cold War biopolitics aimed at fertility. Since ME and MR are registered under different patents, one might therefore conclude that the raw plastic device was made into multiple things at once.[2] The same bit of technology could—by being animated in different assemblages of technique, discourses, and subject positions—be meaningfully said to be two different things. Yet, by attending to the politics of technique—the doing—that animated conflicting feminist and Cold War projects, important *entanglements*, rather than only distinctions, between ME and MR become legible. Entanglements attach distinctive domains as concepts, objects, practices, commodities, and affects pass from one to another, and often back again, in the process undergoing subtle rearrangements by virtue of moving in time and space. These entanglements track the shared conditions of possibility that shaped feminist, nationalist, and imperialist projects. Quite simply, it mattered to any given local use that the device was simultaneously mass produced, commodified, and crisscrossing decolonizing nations. Moreover, and in particular, it mattered to the local instantiation of ME in Los Angeles that the U.S. state sought to distribute MR as part of its foreign aid policy and as an extension of its efforts of "economic development"—efforts that have become an important attribute of its late twentieth-century configurations of empire.

Menstrual Extraction was just one symptom of a tremendous transformation in the biopolitics of reproduction in the late twentieth century. While innovations in the biological sciences offered to manipulate human fertility at microscopic cellular and molecular scales; oral contraception and a plethora of cheap, mass-produced medical devices allowed for new ways of commodifying and governing fertility by individuals and states. Such low-tech developments made possible Cold War projects of the United States and myriad national postcolonial projects that sought to "modernize" national economies by coupling the reduction of aggregate birthrates to the planning of economic productivity. In other words, population control and macroeconomic development schemes produced one another as entangled aspects of late twentieth-century governmentality. The transnational dissemination by the Unites States of millions of units of effective, cheap, mass-produced oral contraceptives in the 1970s was expressive of an emerging liberal Cold War biopolitics seeking to regulate reproduction for the sake of national and transnational economic productivity as well as military strategy. The terms of this transnational *economization of fertility* was an animating condition for feminist practices and protocols that constituted ME as distinct from this emerging dominant governmental formulation.[3]

The economization of fertility names practices emergent in the twentieth century that sought to co-govern national population growth and national macroeconomic growth. Both population and the macroeconomy became national figures that could be counted, measured, and graphed. Measures such as inflation, birthrates, and unemployment rates became governable indexes to be adjusted through national policies. Most important of these was the gross national product (GNP). The GNP per capita, in turn, became a global comparative measure of a nation's productivity, sorting the world into more- or less-developed economies, with more- or less-promising futures. It was a measure that was sensitive to changes in population size, and hence reducing population growth had a direct mathematical effect on its calculation. Through the logics of population control, birthrates, as measures of potential future lives, need to be adjusted to manage a purported postcolonial shift into modernity. Population control policies were thus embraced as part of modernization projects across South and East Asia. Regardless of a state's political ideology, the late twentieth century became a world covered in nations with macroeconomies that could be measured and managed through fertility. Such practices attributed quantitative value to human life relative to macroeconomic growth.

For feminists and Cold War versions of modernization projects alike, reproduction was a pivot on which problems of freedom, control, exchange, and life balanced.[4] While high-tech reproductive technologies have garnered more attention in science and technology studies for understanding the tangle of life, politics, and capital, less glamorous and simpler technologies such as MR/ME have vitally transformed the lives of a vastly greater number of people and have provided a crucial site for the emergence of new practices of governance and new ways of attaching life to economics.

The uneasy entanglements between feminist and Cold War biopolitical projects in the 1970s, moreover, happened at a dynamic moment when many of the contested characteristics of the end of the twentieth century—such as neoliberalism, economic development, NGO-ization, and the governing of population toward the generation of economic value—were producing messy opportunities to experiment with doing life otherwise. As a target of Cold War biopolitics, reproduction was a notably problematized site of experimentation—for feminists, for doctors, for states, for citizens, for activists, and for projects of capital accumulation. Feminists projects were just one of a multiplicity of distinct technical assemblages that struggled over the problem of reproduction. At the same time, feminist biopolitical practices included unproblematized elements within their detailed protocols and techniques, elements which could cut across and connect their projects with hegemonic Cold War possibilities for governing life. Here, I am drawing a distinction between *problematization* as the explicit contestation over or politicization of a biopolitical concern (here reproduction) and *animations* as produced by the entangled and yet unquestioned circulations of ideas, protocols, technologies, and other conditions. What were the unproblematicized terms that moved between ME and MR practice, jointly animating feminist, liberal, and imperial experiments for the contested governing of reproduction? While feminists fashioned ME in opposition to the kinds of population control practices being crafted by agencies like USAID, their opposition nonetheless echoed the emergence of biopolitical practices that reassembled "fertility," "desire,"' experiment," "individual responsibility," and "exchange" as national and transnational concerns.

Protocols from China to Los Angeles

From the late 1950s through the 1960s, China was one of the few countries with an active, state-sponsored research program developing new techniques for birth planning—called *jihua shengyu*.[5] After the Great Leap Forward and its resulting famines in the years before the Cultural Revolution, state interest in birth planning was elevated. As a result, there followed an era of research activity by gynecologists, particularly in the Chung-Hsin Hospital in Shanghai, exploring new methods of abortion (including aspiration but also with electrification and catheters) that could be disseminated into rural China. One concrete invention was a "negative pressure bottle" method of aspiration that used a glass bottle heated with a match to create a vacuum, and hence did not require electricity to create suction. The resulting published work was followed avidly by an emerging group of family-planning doctors, illegal abortionists, and radical feminists who looked to China's abortion research, and later rural "barefoot" medical corps, as examples of how abortion could be provided by paramedical workers instead of by doctors.[6] Untranslated images of this technology clipped from a Chinese nursing journal article circulated among East Coast radical feminists in the late 1960s, repoliticized as a possible means of providing abortions without medical involvement.[7] (See figure 4.3.) Thus, the itinerary of this device begins not with a movement from the West to elsewhere, but from the so-called communist third world to a Western imperial center.

Though Chinese visual instructions were circulated by radical feminists, such as the Redstockings, they were not put into systematic practice. However, a similar device was put into use by Harvey Karman, an illegal abortionist working in Santa Monica. Karman, who himself was not a doctor but rather had degrees in theater and psychology, had been

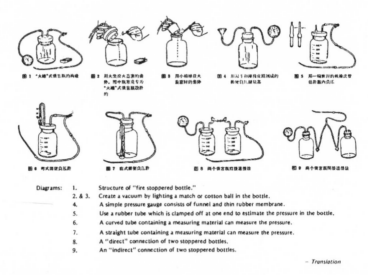

Diagrams:
1. Structure of "fire stoppered bottle."
2. & 3. Create a vacuum by lighting a match or cotton ball in the bottle.
4. A simple pressure gauge consists of funnel and thin rubber membrane.
5. Use a rubber tube which is clamped off at one end to estimate the pressure in the bottle.
6. A curved tube containing a measuring material can measure the pressure.
7. A straight tube containing a measuring material can measure the pressure.
8. A "direct" connection of two stoppered bottles.
9. An "indirect" connection of two stoppered bottles.

– *Translation*

Simple abortion technique, using an ordinary stoppered bottle in which a vacuum has been created by burning alcohol is described in the Peking Journal of Nursing, March 1966. Abortion by suction originated in China where smaller vacuum jars are used to cure headaches and other minor ills.

4.3. Instructions for a manual suction abortion device from a Chinese nursing journal reproduced by the New York radical feminist group Redstockings in their 1975 publication *Feminist Revolution*. Redstockings in the late 1960s and early 1970s were particularly influenced by impressions of Mao Tse-tung's formulation of the Chinese Revolution. The FWHC also republished this image in their *Abortion in a Clinical Setting* (1974).

performing abortions in Southern California for over fifteen years, and already had two arrests under his belt, including one for murder following the death of an abortion patient. He was charming, handsome, and entrepreneurial with a talent for self-promotion. He prided himself on his connections with feminist groups, not only in Los Angeles but also in Chicago, Philadelphia, and France. Moreover, he did not toil in obscurity; he was a well-known figure among physicians prominent in supporting abortion as part of family planning. In fact, Karman was at this moment something like a masculine hero in the abortion scene, giving interviews to the Los Angeles newspapers and even the inaugural issue of *Playgirl*.[8]

In one interview, Karman claimed his eureka moment for MR came to him while in prison.[9] However, suction devices for manually aspirating the uterus had already existed for at least a hundred years. Not only was the U.S. patent office littered with such scattered inventions dating back to the nineteenth century, but non-English published accounts of suction techniques had existed in Eastern Europe since the 1920s. More significantly, information about Chinese suction abortion techniques had traveled widely among Karman's cohort since the late 1960s.[10]

Flexible, sterilizable, reusable, and *plastic*, Karman's design set his equipment off from other versions of aspiration: a simple syringe that could manually create a vacuum combined with a bendable, thin cannula, which he patented as the Karman cannula. Importantly, the cannula tended to bend at, rather than perforate, the wall of the uterus, dramatically increasing the safety of the method and thus inviting its use by nonphysicians. The small size of the cannula allowed it to be introduced into the uterus without dilating the cervix, further simplifying the procedure. The manual vacuum obviated the need for electricity. Plastic was less fragile than glass. In articles by Karman published by medical journals, he suggested his equipment's simple requirements made it feasible for "paraprofessionals" and "paramedics" to employ it in "minimal clinical facilities, particularly in areas where electricity or immediate emergency services are unavailable."[11] Thus, Karman's own formulation of his device rendered it portable, open to nonprofessionals, and usable outside medical infrastructure.

Menstrual Extraction, the Los Angeles feminist self help movement's own iteration of this device, was crafted by Lorraine Rothman, its patent holder. She traces the inspiration for ME to two encounters in the spring of 1971: with Harvey Karman and with Carol Downer, who was then a radical feminist health activist interested in starting up a local underground feminist abortion service along the lines of the now famous Jane project in Chicago.[12] Karman had invited Downer to observe his practice.[13] Karman's manual version of suction abortion was of particular interest to Downer because the unusual technique seemed so simple almost anyone could learn to do it.

Downer was already convinced that abortion techniques in general were not difficult skills to learn, but instead "mystified" by professional medicine. Downer believed that reproduction should be governed neither by the state nor by doctors, rather only by "women." She had apprenticed with the California abortion activist Patricia Maginnis, whose tactics in the late 1960s included circulating pamphlets that taught self-induced abortions using only one's fingers.[14]

Downer introduced Karman's equipment to Rothman, who added a collection jar between the syringe and the cannula. For Rothman, this equip-

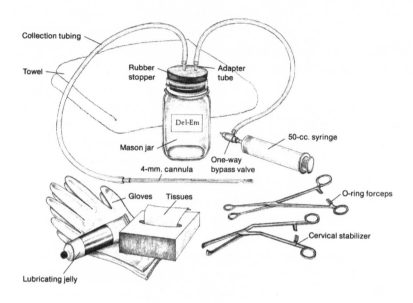

8–1   Menstrual extraction equipment

4.4. Drawing of the patented "Del-em" Menstrual Extraction device. Notice the ordinary commodities of Mason jar, rubber stopper, towel, tissues, lube, and tubing combined with the medical device commodities of the syringe, valve, cannula, and forceps. From FFWHC, *A New View of A Woman's Body* (1991).

ment held the promise that women might learn to safely perform this procedure on each other, making it almost impossible for the state to enforce restrictive laws.[15] It was important to Rothman and her allies that other women who had no prospect of medical abortion—such as in prisons—could build her device with parts found at grocery, hardware, and pet stores (see figure 4.4).

The politics of practice—writ into protocols—was what made Menstrual Extraction distinctive from other kinds of suction abortion techniques for radical feminist health activists in California. According to Rothman, ME "is *not* a medical procedure performed by physicians as a service to women who request an abortion. Menstrual Extraction is *not* a do-it-yourself abortion method. Menstrual Extraction is a new home health care procedure developed by self help clinic women who saw its potential for regaining control of our reproductive lives"[16] (italics added). In many ways, as this quote reveals, ME was constituted by what it was *not*. At the same time, ME necessitated its own assemblage of spaces, practices, and subject positions that (1) staged a moment of exception to law and medical professionalism, (2) emphasized flexible reassemblage and sharing, and (3) encouraged an individually "controlled" sexed embodiment.

Assembled in this way, ME was fashioned as a *noncommodity*—that is, rather than a purchasable technology, it was seen as a mobile, openly available set of operating protocols for assembly. It was not to be mass-manufactured, but cobbled out of other commodities readily available in commercial culture in the United States: canning jars, rubber stoppers, aquarium tubing, even coffee stir sticks. Rather than using Karman's patented cannula, in a pinch a substitute could be fashioned with the plastic tube found in hairsprays, a razor blade, and an iron. Rothman, like Karman, gave no credit to the device's past iterations and was heralded in the movement as a "mother of invention." Precisely because it was not intended for manufacture beyond feminist projects, Rothman patented the ME device as the Del-Em™. The patent described it as an "apparatus whereby substantially all of the menstrual fluid incident to a normal monthly 'period' may be removed."[17]

In a moment when new forms of biotechnologized life were becoming patentable, when property regimes were reaching inward to the micrological substrates of living-being, a moment of what Sarah Franklin has called "biological enclosure" and which ushered in a new regime of ownership of life itself, Rothman secured her patent, not to protect her individual ownership of ME, but to protect her movement's rearticulation of exchange relations as *sharing*.

Patented as a tool, ME was nonetheless conceived as exceeding its physical form. In the words of another practitioner, it "is much more than the technique. It is also the setting, the close community of women, the information sharing."[18] Exchanged only through sharing, the patenting of Menstrual Extraction was meant to secure the ways it avoided commodification. Sharing also formed its experimental method—the experience of each ME was ideally recorded, shared, and used to further improve the procedure. Menstrual Extraction practitioners were at pains to differentiate their project from medical research into contraceptives for the market, particularly research conducted in "areas and countries where people are most defenseless."[19] Promoted in California, the strategy of developing noncommodifiable health techniques through exchange practices of "information sharing" resonated with contemporaneous practices around software development. The exchange economy of ME, conditioned by a new dominant regime of economizing life, was analogous to modes of shared and circulated production that gave birth to software such as UNIX, and later LINUX, as well as the open-source patent, in turn articulated amidst an emerging new regime of intellectual property.[20]

Menstrual Extraction's promise of *doing otherwise* was accompanied by purposively heralding the event as an exception: not only avoiding commodification, it also escaped the law, circumvented professional medicine, and posited an indifference to the condition of pregnancy. Their experimental practice was named "woman controlled research" and was intended to "collect information based on assumptions different than the ones made in medical science and practice."[21] Also important to the staging of alterity was its location outside of clinics: ME took place in homes, amongst secretive groups of feminists. Yet, the exceptionality of ME did not rest on this geographic displacement; it was instead primarily forged through the cumulative effect of its protocols that hoped to stage a radically individualized biopolitics.

As a practice meant to occur only outside law, profession, and commodification, Menstrual Extraction, if not widespread, was iconic of the most radical goals of the movement, the *self-governing of reproduction*, or, as Rothman expressed it, "controlling our own biologies." Menstrual Extraction, Rothman theorized, "places each woman in active control of her period. We no longer wait passively for our monthly visitation. We no longer wait for the first days' cramping to pass. We no longer wait the five to seven days for the whole process to stop. We will no longer accept the denigrating system of myths that a woman's monthly period incapacitates her for several days. . . . Our normal, healthy biological functions are not to be used against us. We choose to have or to not have, when, where and how."[22] By actively grasping the contents of the uterus, ME staged an individualized self-governing that in turn required a regulation of the tempo of sexed living-being. Practiced on the menstrual moment, ME was out of time in relation to abortion. Abortion law did not apply to the first moments of a predicted missed period. A few enthusiasts, such as Rothman and Downer, undertook ME monthly, continuing for several years without a full period. Untimely in relation to pregnancy, ME materialized its object as "menses" indifferent to the presence or absence of a zygote. One handout on ME from the Women's Choice Clinic in Oakland explained, "Her group meets; she and they extract her period, at which point she is not pregnant. Was she or wasn't she? Who cares? She does not; the group does not."[23] In this way, ME secured a state of potential nonpregnancy and the temporal regulation of menses, indifferent to pregnancy itself. If anything ME practitioners sometimes likened it to the IUD as a form of birth control, a technology that physically disallows the bodily securing of pregnancy.

Yet, this ethic of individually governed reproduction was thwarted by a technical fact: it was physically impossible to perform an ME on oneself. Instead, ME had to be practiced in a group in a way that purposively *staged* individualized reproductive control. Within the group, the protocol for ME was carefully orchestrated to grant authorship to the woman having the extraction (see figure 4.5). She directed the actions of the others. She inserted the speculum into herself and pumped the syringes to create a vacuum in the mason jar. Another group member would insert the cannula into the opening of the cervix, all the while narrating her actions and discussing them with the others. The woman receiving the extraction would narrate her sensations as well and direct when the procedures should start, stop, or change pace. Narrative scripts emphasized sentiments and sensations. Another participant took notes. The contents extracted, either menses or an early pregnancy, were examined afterward. In the utopian form of ME, all women would have such a group to call upon around the time of their period.

The individualized sovereignty of a woman over herself, was understood to only be possible through this mobile, flexible protocol that assumed the larger aggregate of "women," smoothed of race, class, or location, as its field of circulation. Advocates of ME imagined a universalizable ME that served the individually determined interests of any and all women. Feminist self help underlined that it was a woman's individual choice whether to have or not have a child, at the same time emphasizing that it took specific techniques to maximize the freedom of this choosing. Included in these techniques were social scripts that hailed women in general as ethical subjects who could be emotionally bonded to one another through sentiment. What characterized the biopolitics of ME was thus also a fostering of sociality induced though affects such as bonding, joy, exhilaration that posited an emotional and politically linked, yet unraced, sisterhood.

Thus, ME was a biopolitical project simultaneously on a microscale and macroscale. The microbiopolitical effort sought to technically create indi-

8–3 A woman inserting her speculum

8–4 The woman who is having the extraction pumping
the Del-Em

4.5. Illustration by Suzanne Gage of an ME within the small group format.
Here the woman receiving the ME is shown inserting her own speculum
and creating the vacuum in the syringe, thus situating her as in charge
of the procedure. From FFWHC, A New View of A Woman's Body (1991).

vidualized control over the sexed reproductive body, while the macrobio-political register was traced by the flexible, universalizable, and mobile features of a protocol intended to bind and circulate among women in aggregate anywhere and everywhere. Oriented at both micropolitical technique and unlocalizable macropolitical women, the project of ME had little to do with the terrain of local urban reproductive politics in California. For example, the Los Angeles feminist self help movement literature makes no reference at all to simultaneous struggles over nonconsensual hospital sterilization that Los Angeles Chicana feminists were waging in the courts and streets during the very same years. Local and raced reproductive politics, in fact, were almost never the horizon in which practitioners of ME understood their biopolitical projects (unlike, for example, community health clinics). Instead, ME was resoundingly formulated as a challenge to the emerging population control industry in the United States, and it was profoundly constituted through its antagonism with MR. It was staged as MR's other. This antagonism nonetheless traced common, even if disputed, animating terms for the practice of "freedom through fertility."

Supplying Cold War Contours

Harvey Karman not only showed his device to local feminists; he hawked it amongst a constellation of doctors, researchers, and policy makers brought together through the emerging project of family planning in the Cold War. Karman successfully enrolled the interest of Malcolm Potts, the first medical director of International Planned Parenthood Federation (IPPF), and of Reimert Ravenholt, the director of the then five-year-old Office of Population at USAID, which had just been allotted a financial windfall, ushering in an era of intensive investment in Cold War family-planning projects that lasted until the Reagan administration.

Karman's device fit into USAID's newly formed "inundation strategy," or "supply-side" strategy, of the late 1960s and early 1970s in which condoms and oral contraceptives were "donated" to national family-planning projects in "least developed countries" of the "free world." Supply side was an emergent neoliberal strategy that coexisted among a constellation of strategies—some directly coercive—for governing fertility that were evolving at this historical moment. Ravenholt, for example, rejected population alarmists claims that the exploding population growth of the poorest parts of the world could only be curbed with coercive measures.

He also argued against the view that economic development was a necessary prior condition to women's choosing to reduce their fertility. Nonetheless, supply-side strategy had its own economic logic: if supply is not there, no demand will exist, but if supply is abundant and easy to reach, demand will be high.

Ravenholt likened his logic to the way Coca-Cola was sold as a global commodity. The importance of the supply-side principle was "demonstrated with Coca Cola: If one established a few places in a country where Coca Cola could be purchased, at a considerable distance from where most people live, at uncertain hours, necessitating that buyers have a sophisticated knowledge of the distribution system, no doubt the sale and use of Coca Cola would correlate considerably with general education, occupation, economic and transportation circumstances, sophistication, etc. But if one distributed an ample free supply of Coca Cola into every household, would not poor and illiterate peasants drink just as much Coca Cola as the rich and literate urban residents?"[24]

Similarly, changing material conditions by simply presenting women with a freely given commodity—contraception—would provoke an "unmet need" that the contraception would then fill. Though women were not required to pay for contraception themselves, they were called upon to inhabit the site of consumption in circuits that funneled money, for example, from USAID to pharmaceutical or medical device companies that supplied products and to distributing NGOs and state agencies that supplied services. With its supply-side strategy, USAID quickly became the largest distributor of contraception to the world, minting 780 million monthly cycles of their own Blue Lady brand by 1979. The Blue Lady brand used a standard nonproprietary packaging that featured a racially neutral blue woman taking the pill. Having their own brand allowed USAID to contract with any pharmaceutical company without "confusing" users. Some 2 billion cycles of Blue Lady Oral Contraceptives were circulated over thirty years, purchased in bulk by USAID for as little as fifteen cents, making this one the world's most widely distributed oral contraceptive brands in the early 1970s.

Supply-side strategy preferred pills to IUDs.[25] Not only did pills require no investment in health infrastructure; they were thought to be more attractive to women, called "acceptors," who were more likely to feel that they were "doing whatever they wished to do."[26] The idea, as implemented in the world's longest-running experimental field site for fertility research—in Matlab Thana, Bangladesh—was that pills and condoms were not just supplied to doctors or even clinics, but dispersed by lay "Lady Village Workers" going door to door as "motivators." By circumventing the need for medical infrastructure or doctors, the distribution of pills and condoms door to door would rapidly "light a contraceptive fire" (see figure 4.6). Reflecting back on his policy later in life, Ravenholt commented, "We learned in Egypt, Bangladesh and many other countries, that [oral contraceptives] could be rapidly introduced into the poorest populations, simply by distributing 3 or more cycles to every household willing to accept such a packet from visiting field workers."[27] Supply side enunciated an emerging neoliberal governmentality in which the right market practices would result in "even the peasant woman" choosing to consume birth control, thereby accomplishing "the enlargement of human freedom by extension of family-planning programs."[28]

In Bangladesh and elsewhere, Ravenholt and family-planning professionals supplemented this family-planning project with the collection of data through demographic and attitude surveys that helped to craft a new target of governance: "unmet need." The concept of unmet need was defined as the measurable gap between women's aggregate desire to control their fertility and the unavailability of contraception. Ravenholt's supply-side approach reframed the concept. He argued that instilling "need" into women was itself *relative* to the availability of contraception, putting a contingency twist into the query "What do women want?" Offering contraception triggered new desires for it.

Ravenholt explicated his supply-side strategy and its unmet need to the first USAID population conference by using an experiment with the audience. At the coffee break, he had secretaries give the north half of the audience a questionnaire about their desires, and offered cookies to the south half. The questionnaire read as follows:

Dear Population Conference Participants:
Anonymous December 8, 1976
Age: Number of Living Children
We wish to ascertain your priorities. Please check the items
below you would wish to have at this time.

  1. An apple   4. Cake
  2. Coffee   5. Cookies
  3. Tea   6. Candy

CONTRACEPTIVE   AVAILABILITY

The Key to Rapid Utilization

COUNTRY
SUPPLY

CLINIC
AVAILABILITY

VILLAGE
AVAILABILITY

HOUSEHOLD
AVAILABILITY

4.6. An image that Reimert Ravenholt, director of USAID's Office of Population from 1966 to 1979, used to illustrate the so-called supply-side strategy, which emphasized contraceptive availability as the key to rapid uptake of family planning. Here, contraceptive availability is the crucial ingredient to lighting a family-planning fire. Contraceptives distributed door to door become the kindling for rapid ignition. From Ravenholt, "Taking Contraceptives to the World's Poor: Creation of USAID's Population / Family Planning Program, 1965–1980." Courtesy of Reimert Ravonholt.

Ravenholt later tabulated, "Of the 63 persons in the north half of the room who completed questionnaires, 8 (13%) indicated they would like a cookie; whereas of the 83 persons in the south half of the room who filed out the south door and were offered cookies, 80 (96%) took one or more cookies."[29] The audience was thereby instructed in how desire followed after availability. Need and desire were sentiments brought out by technically altered circumstances of consumption. The literature from USAID, in turn, developed a visual representational trope of an emotive moment of desire-fulfilled in its images of women accepting contraception.

Thus, in this incipient neoliberal Cold War biopolitics—with the help of coke, cookies, and campfires—USAID crafted an assemblage of personal choice, commodity, and unmet need in which marketing logics maximized freedom at the target of fertility. Through the staged opportunity for contraception, an imagined exchangeable rural "third world woman"— for which the residents of Matlab Thana became an important stand-in— was called upon to name and then fulfill instilled desires. Put another way, the supply-side strategy aspired to call forth unmet need, a primarily psychological entity, make it manifest through a demographic-marketing-commodity assemblage, and then operationalize it through a form of governance that asked women to enunciate their freedom and desire in terms of consumption and fertility.

Menstrual Regulation, like the pill, seemed to hold the promise that it too could be a commodity disseminated without the need to invest in medical infrastructure or highly trained personnel, or even the provision of electricity or clean water. The USAID Office of Population contracted to develop a disposable MR kit (see figure 4.7) for cheap mass production. At $8.70 each, USAID ordered ten thousand kits to be distributed to the approximately three hundred practitioners and policy makers from over fifty countries invited to a USAID conference on Menstrual Regulation in Hawaii. With this success, USAID ordered 100,000 more. Coverage of the conference in *The New York Times* reported, "It is something we will be able to bring practically into the rice paddy."[30] The specter of USAID distributing abortion kits quickly attracted Christian senators, instigating the Helms Amendment of the Foreign Assistance Act in 1973, which today still forbids federal funding of abortion and of involuntary sterilization, and thus put an end to USAID's direct distribution of the kit.

Yet Karman had also won over Malcolm Potts's interest in his device.

STERILE: contents of inner wrap are sterile unless it has been opened or damaged.

menstrual regulation kit

• for single patient use
• for use only within the first 40 days of amenorrhea

contents:
- **sani-spec®** vaginal speculum
- lubricating jelly (5 grams)
- 60cc vacuum syringe (locking plunger)
- 3 absorbent gauze sponges (u.s.p. type VII)
- 4, 5, & 6 mm cannulae (1 ea, karman type *)
- specimen cup
- tampon
- 2 gloves
- instructions

CAUTION: Federal (USA) law restricts this device to sale by or on the order of a physician.

burnett instruments¹
700 E. 22nd St.
Lawrence, Kan. 66044

A *licensed under U.S. Patent 3,769,980   **div. of C.R. BARD, Inc.

4.7. Disposable Menstrual Regulation kit manufactured by Burnett instruments and distributed through USAID-funded organizations in the 1970s. From Richard Soderstrom, "Menstrual Regulation Technology" (1979).

Through the IPPF, Potts called on Karman's skills to perform abortions some eight thousand miles away from Santa Monica, at the aftermath of the violent war of 1971 that saw the separation of West and East Pakistan, and the founding of Bangladesh. By the end of nine months many were dead, and some millions of people became refugees. Moreover, national-ist sexual violence had been systematically deployed, resulting in large numbers of raped women.[31] In the recovery, the newly formed Bangladesh government declared raped women heroines of war, *birangona*, and estab-lished rehabilitation centers to provide medical care and other assistance. Nonetheless, pregnancies from war rape were widely seen as polluting the distinctiveness of the new nation.[32] The state temporarily suspended the illegality of abortion. The IPPF was called upon to set up abortion clin-ics at rehabilitation centers, bringing in abortionists from the United States, England, and Australia to offer medicalized abortions to raped women.[33] Among the abortionists were Harvey Karman, Malcolm Potts, and Leonard Laufe, an obstetrician who would later found an NGO called IPAS that would later manufacture MR.

The importance of NGOs to MR was characteristic of USAID's approach to fertility control, which knitted together reproduction, economy, and experts in a way that harbingered later neoliberal forms of development.

Instead of directly carrying out family-planning programs, USAID funded nongovernmental agencies and universities to conduct research and perform projects on the ground. The Carolina Population Center in the Research Triangle, initially funded by USAID and the Ford Foundation in the mid-1960s, was indicative of just one local efflorescence of NGO-ized research. It cultivated a cluster of what are now considered NGOs, among them the International Fertility Research Program (IFRP)—funded in 1971 through another USAID grant in order to "accelerate the development and testing of new and improved means of fertility control for worldwide use"—and which coordinated overseas clinical trials for MR and other devices. Also, there was the International Pregnancy Advisory Service (now only known as IPAS), which manufactured MR devices and was a direct reaction by Potts, Ravenholt, Laufe, and other similarly networked family-planning experts to the Helms Amendment, allowing MR to continue to be manufactured and funded at arm's length from the state and the university. The IPPF in turn supplied the overseas clinics where the experimental device could be used and data collected. A research infrastructure was crystallizing in which the clinical trials for new fertility control devices or drugs were joined to their distribution through transnational NGOs. In other words, *experimentation* was a constitutive component of the assemblage that made up the transnational biopolitical targeting of fertility through MR.

Thus, despite the Helms Amendment, in 1978 some 175,000 kits had been distributed with the help of the IPPF, the IFRP, and IPAS, through which some five million procedures had been performed since the Hawaii conference, retroactively declared a "clinical trial."[34] The most noted MR success story of this involution of aid and experiment was again in Bangladesh, where U.S. funding, circulated through the Pathfinder Fund, helped develop a national program that included the performance of MR by Lady Village Workers, sometimes employing *birangona*.[35] Again MR designated an exception to the illegality of abortion, legalized in Bangladesh as a form of nonabortion, an "interim method to establish nonpregnancy."[36]

Cracking open the USAID assemblage animating MR, a distinct abstract algorithm can be found to haunt the twenty-first century's development-experiment complex that today seeks to maximize economic efficiencies as "freedom." Outside of capital investment in medical infrastructure, at the interstices of law, laypeople are enjoined to become participant-experts in their own life management and in so doing to also become experimental subjects in their own individualized survival as well as in transnational experiments. In practice, USAID was quite willing to fund national population-planning projects that did not have such lofty aims, such as those in 1980s Bangladesh that directly used coercion to reach national birth reduction targets. Seizing the means of reproduction, then, situated feminist projects in a complex Cold War experimental field—in which governmentality, reproduction, and experiment interweave not only in the clinic, but in economic development projects, demographic field sites, population policies, and NGOs.

While this constellation of NGOs was funded directly or indirectly by USAID, their tactical distance from the state allowed researchers and family-planning programs to perform in a less regulated environment, and at the same time foster new health services as a privatized, even when not-for-profit, domain. Moreover, it allowed the emerging family-planning industry to take advantage of the gaps produced by the uneven regulation of experimentation across nation-states, such that the site of aid—a clinic or a field site—could become simultaneously a site of experiment. Thus, with the help of this conjunction of actors, family planning generated an infrastructure that bound the effusive circulation of commodities, with the creation of experimental value, with the provision of a service rendered as planet saving as well as individually liberating and lifesaving.

The Cold War development of MR, therefore, sits in a direct genealogical relationship with current iterations of biocapital in what is now called the transnational Clinical Research Organization (CRO).[37] The International Fertility Research Program (FHI)—as an incipient CRO—became the nonprofit Family Health International, which today is the largest recipient of U.S. federal funds to conduct transnational HIV research, and is among the largest transnational NGOs working at this intersection of research and aid. In 2005, for example, the FHI received over $100 million in federal President's Emergency Plan for AIDS Relief (PEPFAR) funds. The clinical research networks that could both disperse family-planning devices and assemble them as experiments were transformed into CRO platforms that could deliver to the state HIV research, on the one hand, and deliver to corporations, such as Coca-Cola, prepackaged humanitarian endeavors, on the other. The networks created by FHI, in turn, spawned a series of for-profit CROs, including PharmalinkFHI in 1998, the first clinical research organization to manage transnational clinical trials through the Internet—called an eCRO.[38] The chairman of FHI, Albert Siemens, has simultaneously held numerous executive positions in many of these spin-off CROs, which have successively been sold off, renamed, and merged. The infrastructure spawned by the family-planning project funded by the United States became a platform for the growth of transnational biomedical experimentality.[39]

Refractions

The device's ease of use outside medical infrastructures—by laypeople, by feminists, and in resource-poor rural settings—underscores the uneasy entanglements across biopolitical and geopolitical difference. Yet feminist self helpers did not just see ME as a localized alternative to MR's population control circuits. That is, ME was not just an alternative to an imperial formation; it was imagined as, what I'll call, a *counter-empire project*, a project that contests and yet still partially reanimates empire. In its most arrogant moments, feminist self help aimed to replace population control everywhere. Like MR, ME could be fashioned as a mobile technology, not because it worked immutably no matter the circumstances, but because it hailed a transposable and flexible ethical subject at its center. Menstrual Extraction, however, was not a commodity. It was a technique to be "shared," not consumed. To make ME mobile, feminist self help created illustrated manuals and pamphlets; its most formal, *When Birth Control Fails* (1975), was instigated by a request from the feminist journalist Barbara Ehrenreich, who asked Rothman to come up with a manual women imprisoned in Chile might use. Ironically, the manual was in English, while detailed illustrations were meant to do the work of translation. Nonetheless, ME instructions circulated widely. The self help manual, not the instrument itself, was the vehicle for making ME portable.

Feminist self helpers cast themselves as the mirror opposites to Ravenholt and other professional luminaries of the budding international family-planning industry, which they carefully tracked. Their literature contains exposés and photographs of their confrontations with well-known family-planning figures. Going to UN-sponsored and other international conferences allowed ME advocates to envision themselves on a "global" terrain. For example, Laura Brown, the director of the Oakland Feminist Health Clinic attended USAID's Hawaii conference, recounting that she was grudgingly received and followed by a "public relations man" for the four days of the conference. According to Brown, her attendance was differentiated from that of other women at the conference because they were paid by USAID to participate in and proselytize MR. "Naturally," one feminist commentator reported, "Laura was able, as a laywoman, to interact much more effectively with Third World Women."[40] Belief in solidarity of women everywhere failed to recognize non-Western women as nonidentical persons, or already as feminists; rather, they were perceived as masses in need. I have not found a single person from the global South individually named in the ME literature despite the literature's global imaginary.

For Hirsch, Downer, and others, Bangladesh stood out in their imaginary as an exemplary "third world" site where undifferentiated "poor women" were experimented on. AS an experimental practice, ME was juxtaposed with what might now be called a postcolonial global research economy. Menstrual extraction did not experiment on others, according to Rothman, but "use[d] our own bodies" and allowed participants to "choose which risks are acceptable for each one of us individually."[41] Interestingly, the example of Bangladesh was also used to reverse the normative grain of colonial difference. In Bangladesh, it was argued, USAID and IPPF had proven how nonprofessionals could perform abortions, yet here in the United States, it was argued, "we are terrorized into believing that only highly trained professionally educated men have the brains and the expertise to learn how to do an abortion. We are intimidated into believing that laywomen in the United States are too dumb, too uneducated, too dirty, too infection-prone, too inept, too unmechanical, too averse to technology to learn how to do an abortion."[42] The same device in the United States could only be legally used by doctors, while in Bangladesh, under U.S. state encouragement, it could be used by trained laywomen, thus marking a boundary of uneven biopolitical distributions.

The counter-empire of feminist self help was perhaps most clearly enunciated in the words of one clinic director as "a *global women power which transcends the nations of men.*"[43] Menstrual Extraction was to replace Menstrual Regulation, and the feminist self helper was to replace the family-planning professional. Following the MR promotional conference in Hawaii, feminist self helpers held their own promotional conference in Oakland, where they demonstrated ME on themselves for the audience. At the subsequent second feminist self help conference held in remote Ames, Iowa, the participants temporarily interpellated themselves at the center of the world, as "the most important 175 people in the world population of 6 billion (11,000 added ever hour and 95 million added every year)."[44] The movement's small, mimeographed newsletter, the *Monthly Extract*, heralded itself as a "global Communication Network," though it managed to stretch across the Atlantic only as far as Germany. Feminist self help was inside an imperial vantage point that it marked as out there, not here.

Menstrual Extraction defined by its foil MR, moreover, had common constitutive elements, among them (1) need fulfilled with choice, (2) fertility as a necessary route to freedom, (3) experiment folded into practice, and (4) a technical fix that worked flexibly in the field. Menstrual Extraction and USAID's version of MR actively sought to create the conditions in which individuals could be called upon to "choose" to govern themselves. *Freedom, for the idealized versions of both ME and MR, was not an absolute withdrawal of governance from the domain of reproduction. It was a rearrangement and reinvestment in techniques that created the conditions generative of managing one's own fertility.* Women were to be "responsibilized" to make choices about their own fertility.[45]

Yet for radical feminists, the freedom of choosing happened in a moment of micropolitical staged exception (even if complicated by their vision of "women" as a global aggregate), while for USAID the freedom of choosing happened in a moment of enrolling women into a relation with global capital. One stressed freedom via control as exercised in techniques of self-governing one's biology at the scale of the individual, outside the grasp of state or medical regulations. The other operated through nation-states, multilateral agencies, and NGOs to curb and govern population growth at the transnational scale, preferably with technologies that acted as immutable mobiles, that is, which operated consistently no matter the circumstances of delivery.[46] The feminist self help movement's central aim was to help women become expert at controlling their own reproduction in a way that maximized one's freedom in the moment of health care — freedom from state law, and from reliance on medical expertise, as well as freedom to control one's own biological being: the "means to responsibly control our periods."[47] ME was meant to free the control of reproduction from other biopolitical interests: "the results suit our needs, not the needs of doctors, capitalists and population controllers."[48] As problematizations of reproductive health, ME and MR were indeed distinct technological instantiations that mattered differently.

The tracking of entanglements between feminist self help and family planning is not intended to belittle, ignore, or level the tremendous differences that characterize their history nor their rivalry that have had consequences on lives. Entanglements should nevertheless be seriously considered, since it may help explain some of the persistent convergences between feminist projects and population control projects in the late twentieth century, particularly the ease in which they passed into each other and blurred. Feminist self helpers themselves were deeply disturbed by these confluences as evidenced by their constant gestures to distinguish ME and MR.

The exchange of protocols, the provision of service by laywomen, and the goal of supporting the self-governing of reproductive health by *individuals* all became wider standards of care in neoliberal relations of rescue support by the United States. Nongovernmental organizations, many of them feminist, became the primary recipients of funding for the provision of women's health services, and in turn feminism itself became NGO-ized. This was something radical feminists of the early 1970s certainly did not imagine. In short, NGO-ized feminist health care became the *model* for what population control programs should look like in the field. This was the case with the provision of MR in Bangladesh. With support from the Population Council and assistance from the International Women's Health Coalition, a New York–based transnational feminist NGO (discussed in chapter 3), the Bangladesh Women's Health Coalition (BWHC) began as a feminist MR clinic in 1980. Arguing that a feminist version of reproductive health care in a Dhaka slum was not an "unobtainable luxury," BWHC describes itself as offering a "flexible, woman-centered reproductive health program responsive to women's needs" that goes beyond MR to include children's health care as over one-third of its services.[49] In the 1990s, BWHC was held up as a transnational model of feminist reproductive health, performing a feminist appropriation of MR that could circumvent the logics of both state population control and USAID funding restrictions.

Thus, ME was not always so radically different from MR. Through versions of the self help manual, many NGOs with a feminist twist formed a convoluted chorus calling to a global female subject who for the sake of the planet, for the sake of the nation, for the sake of preventing war, for the sake of undoing disparity, and for the sake of her own liberation, was

enjoined to be an ethical subject who facilitated the control of her reproduction through appropriations of many kinds.

Here and There

The history of entanglements between ME and MR offers several important methodological and political lessons. The first is that transnational
history is not merely a matter of scaling up. Local places and small-scale
acts are shaped by larger events—in this case Cold War and international
family planning, among others. The transnational is present in the "here"
of little actions and daily decisions. The use of a particular instrument,
such as a suctioning syringe, is shaped by the broader circulations of protocols and other suction abortion devices without itself having to travel;
for that reason I have argued that the history of reproductive techniques
has been profoundly shaped by *animations* and not just explicit entanglements or literal exchanges.

The second lesson is that the histories of critical political projects are
not severed from the phenomena they critique. More specifically, feminist and nonprofit health projects—not just state, corporate, and institutional projects—have been caught up in the history of neoliberal practices. Moreover, through entanglements, feminist practices have helped
to *render ethical* (that is, both inject ethics into, and help to legitimate)
some of the constituent elements that have shaped neoliberal family planning and public health.

If feminist projects are caught up in entanglements, does this analytical
formulation offer insights of strategic importance for the ongoing project
of striving for better versions of technoscience and feminism? Foucault
characterized biopolitics as the regulation of life and the management of
letting die. In the late twentieth century, biopolitics was pushed into the
terrain of the prevention of birth—a promissory terrain of potential life,
potential futures, potential profits before it is possible for death to arrive.
The anticipatory politics of future human life penetrated bodies, moved
through technologies, and reached into national economic planning. Our
understanding of the history of twentieth-century biopolitics necessarily
becomes altered when it is spatialized and multiplied, and not just tem-
poralized—that is, when entangled biopolitical topologies are mapped.
The story of ME and MR, then, suggests that the ongoing feminist project

of practicing technoscience for the sake of a better way of living is not containable to clinics or bodies or rights. Feminist practices of care articulated in Los Angeles were noninnocently animated by (though importantly not fully subsumed by) larger globe-stretching projects with nodes in sites such as North Carolina, Washington, D.C., and Matlab Thana. The entwined transnational histories of ME and MR materially set the stage for the further development of a global industry of drug testing and transnational HIV research infrastructures. Together feminist and family-planning projects encouraged a sensibility toward an individualized ethical subject who was personally accountable to the management of life, health, and futures, as well as national prosperity. In concrete and practical ways, then, feminist biopolitical projects have been both urgently necessary and always noninnocently imbricated.

Living the Contradiction

What does it mean to "seize the means of reproduction" when reproduction is so exquisitely bound up, not only with living-being, but with economic and governmental practices? In attempting to answer, this book has mapped some of the shared but agonistic relationships between feminisms and technoscientific practices over questions of reproductive health that took place during a period marked by Cold War and postcolonial transformations in the governing of life. By following technologies and practices as they moved in and out of feminisms through clinical, urban, national, and transnational terrains, I have been motivated as much by trying to understand the conditions of possibility for reproductive politics in the present, as by the need to better excavate the past. I have tried less to capture a social history of what it was like to be a feminist fashioning a new kind of reproductive politics in Los Angeles, and have concentrated instead on mapping some of the tensions within reproductive health politics as it was assembled during the last decades of the twentieth century.

The late twentieth century, I have argued, was populated by transnational and local experiments in governing reproduction and health that have since crystallized into pervasive and familiar contemporary practices—from cost-benefit analysis to participatory research—concerned with women's health both in the United States and elsewhere. While I set out to learn from the counter-conduct of past feminist health projects, I quickly found myself also grappling with the contradictions that situated feminisms as participants in the histories of neoliberalism, postcolonialism, racial governmentality, and American empire. With an eye to the larger historical stakes, I have argued that feminisms can offer a critical diagnosis of their moment and are at the same time symptomatic of the conditions of their articulation. Methodologically, understanding feminisms as both diagnostic and symptomatic has involved mapping the wonders and the horrors, the troubling and the inspiring—that is, the discrepant and the synergistic ways feminisms and technoscience were brought into being together and did contradictory work in the world. Feminisms were an important form of biopolitical counter-conduct in Cold War and postcolonial histories, and hence are as useful an entry point into the larger history of biopolitics as more typical topics, such as "big science," genomics, or particular diseases.

The histories of circulating epistemologies, practices, and technologies told in this book were accompanied analytically by important conceptual terms for rethinking the histories of feminisms and technoscience. First, the book argues that the late twentieth century saw the emergence of *protocol feminism*, a kind of feminism that posited its politics in the technical details of practices. Protocol feminism is an example of the "politics of politics," that is, the historically specific means by which some domains and not others are understood as political. Second, the book holds that feminist self help crafted a moral economy of *immodest witnessing* in its research, in which it was crucial that the subjects of knowledge making were embodied and acknowledged. They insisted that objectivity was improved by considering the question of subjectivity, not in abstract philosophical terms, but instead as an expression of particular embodiments with specific identities. In doing so, they exemplified an important episode in the twentieth-century history of objectivity—that is, the emergence of the belief that objectivity is better achieved if the people making knowledge consider their own embodied and situated subject-positions as starting points.

Methodologically, I hope that a significant contribution of this work is its effort to historicize biopolitics as a panoply of situated phenomena. The book attends to the work of scale—from individual bodies, to clinics, to national policy, to transnational public health—in the making of multiple biopolitical orientations. It shows how the scale of historically available and imaginable solutions molded the very ways health problems have been rendered. Moreover, the book shows how local instantiations of health and reproductive politics within the United States were deeply shaped by and entangled in larger transnational histories. Its methodological approach to both these questions of spatialization has been to conduct research that follows technologies and practices as they travel between different sites and inhabit different scales. As a result, the book offers an example of how one might spatialize the history of biopolitics, and therefore tell stories that provincialize histories of American feminisms.

Today feminisms can be found at sites as diverse as the World Bank headquarters and street protests, making attempts to discern kinds of feminisms inadequate. Feminisms, as a multitude, are not merely a taxonomy of ideological kinds; they can be profoundly antagonistic to one another. Feminisms are assembled and performed in myriad actions, details, and practices that cannot hope to fully escape larger historical forces. To study feminisms and technoscience as entangled is to offer an account that charts generative, and not just divisive, contradictory relations. An implicit claim of this book is that proliferating feminisms in the late twentieth century were assembled by virtue of borrowing, remaking, sharing, and appropriating other practices and epistemologies. In turn, feminist practices are themselves regularly appropriated. Thus, the critical question becomes not whether an analysis is for or against feminism; rather, how is feminism performed here, through what reassemblies, and to what contradictory effects?

"Living the contradiction" is a phrase colloquially used to describe the constitutive incoherence and agonistic elements that form daily life for a self-consciously politicized actor. When the largest feminist health NGO based in the United States works closely with diamond-mining corporations, it is constituted in a state of contradiction. Feminist theorists have in an abstract way accepted the impurity of all political and analytical projects, and this book therefore aspired to excavate the intensive landscape of entanglements that have become the ground of contemporary protocol feminism and the women's health movement. At its extreme, contemporary feminist health projects are less characterizable by ideology or moral economies than through their proliferation of mobile procedures designed for insertion into both dominant infrastructures and local activities.

"Living the contradiction" also describes the very ontological politics of sexed living-being in the shadow of American empire. From "ontological collectivities," to "immodest witnesses," to the "not uncommon," to histories within histories, this book has been concerned with not only the supportive "ands" that make up assemblages but with agitated and contradictory "ands" that uncomfortably entangle sexed living-being as well as cleave. Living the contradiction is the collective condition of feminisms, technoscience, and reproduction within an uneven, paradoxical, and shifting biopolitical topology.

What, then, does "seizing the means of reproduction" mean as a politics if reproduction is entangled in such convoluted, layered, and geographically extensive ways? The challenge is to remake yet again the ontological politics of reproduction—of generative sexed-life bound not only to race and capital, but accountable to affect, death, and human-ness itself. Feminists wielding speculums, performing Pap smears, and offering menstrual extractions were indeed not just practicing technical acts; they were rematerializing sexed living-being as vaginal ecologies, as lush fields of variation, and as biologies under the authority of sovereign selves. They were engaging in ontological politics—the contested, power-laden materializations of living-being in the world. The ontological politics of reproduction is not settled. What reproduction might become remains open, both because it is a process open to technical alteration and because of the possibilities of conceptually reimaging the scope and extension of reproduction, and hence what *reproductive politics* might refer to.

An example of the ongoing remaking of reproduction is found in the notion of "reproductive justice." This term was fashioned in the first decade of the twenty-first century by activists associated with the Sister Song Women's Reproductive Health Collective, itself composed of many women-of-color feminist groups from across the United States. *Reproductive justice* was defined as "the complete physical, mental, spiritual, political, economic, and social well-being of women and girls"; it "will be achieved," according to activists, "when women and girls have the economic, social and political power and resources to make healthy decisions about our bodies, sexuality and reproduction for ourselves, our families and our communities in all areas of our lives."[1] Activists used this definition to expand the terrain on which reproduction was at stake, including such issues as militarism and environmental degradation as practices within the scope of reproductive politics. Reproduction within the bounds of bodies, in this reframing, was actively shaped by a host of structural inequalities that exceeded the scope of medicine.

At stake in the histories offered in this book is not just how reproduction is politicized, but what reproduction is: how far from bodies it extends, and how it binds people together, hence the scales of political invention required to secure the flourishing of present and future life. In this sense, reproduction is not self-evidently a capacity located in sexed bodies. Reproduction has been a multiply rendered distributed formation, joining cells, protocols, bodies, nation, capital, economics, freedom, and affect as much as sex and women into its sprawl. Reproduction, as multiply made in a complex biopolitical topology, joins men, children, lives unborn, economists, soldiers, and lawmakers into its ambit in ways that put feminism as the moniker for this politics into question.[2] The question of how to "seize the means of reproduction" is thus reversed. Practices do not just capture reproduction into their ambit; how we constitute reproduction shapes how it can be imagined, altered, and politicized. Perhaps here, a double vision is not enough to track the entanglements that connect in the same gesture through which they sever. Feminism is necessary, but cannot alone hope to be sufficient to the task of reassembling reproductive politics as life within the contradictions.

Introduction

1. One example here is women assuming the responsibility in practices of prenatal diagnosis, both in terms of their position as the subject who must consent to the procedure and in terms of the responsibility to decide whether to terminate a pregnancy. This particular example has been called a moment of "flexible eugenics" in which decisions about reproduction, and particularly whether to keep or terminate a pregnancy, are morally and technically configured as "technologies of the self" inhabited by individual women. See, Rayna Rapp, Deborah Health, and Karen-Sue Taussig, "Flexible Eugenics: Technologies of the Self in the Age of Genetics."

2. The women's health movement, like the notion of the women's movement more broadly, might be thought of as an imagined community, an interpellation of common cause that is tactical, interpellative, and at the same time masks the fissures and profound contradictions that make and shape the diversity of feminism. The term *women's health movement* is used to hold together feminist health centers, feminist physicians and policy makers, feminist NGOs, feminist reproductive politics of many kinds, and so on that not only vary among themselves in terms of the political ideologies and tactics, but also take different shapes in relation to local, national, and colonial formations of race, class, sex, and capital. Thus, I'm trying to attend to the disunity of feminism and largely avoid using the broad term *women's health movement* in the book. On the history of the women's health movement in the United States, see Sandra Morgen, *Into Our Own Hands: The Women's Health Movement in the United States.*

3. I'm using the notion of counter-conduct in this book, rather than, for example, resistance. Resistance has acquired a romantic moral valence in Left academic work as a self-evidently desirable set of actions antagonistic to hegemony. Counter-conduct, in contrast, invites a historicization that highlights modes of undoing, remaking, and antagonism that are immanent with and animated by hegemonic formations. Foucault used *counter-contact* in his lectures of 1978 to describe actions that deterritorialize hegemonic governmentalities; Arnold Davidson has further developed the term. *Conduct*, in English, is a rich word which both allows one to think of nonhege-

monic *practices*, *behaviors*, and *directives* as well as to think of conduction as a path or strategy of entanglement and movement. I use *counter-conduct* in a broad sense, not just to refer to arrangements of behavior and subjectivity, but also arrangements of exchange, affect, and technical practice. I am using the term *counter-conduct*, moreover, not to name a line of flight that is an absolute deterrorializaiton, but rather as an immanent unmaking that is also simultaneously a remaking of another minor, or nonhegemonic formation of conduct that remains conditioned by and entangled with the hegemonic. Finally, I've also tried to see the emergence of feminist biopolitics as a counter-conduct in terms of its multiplicity, spatialities, and entanglements, rather than as a singular overarching reaction. In a more abstract register, I am trying to imbue the Foucauldian notion of counter-conduct with a more complex sense of becoming aided by both feminist and queer theory and the work of Gilles Deleuze and Félix Guattari.

4. On experimental systems and cultures of experiment in the life sciences, see Hans-Jörg Rheinberger, *Toward a History of Epistemic Things: Synthesizing Proteins in the Test Tube*. On the experimental logos in colonial and postcolonial governmentality and biomedicine, see Christophe Bonneuil, "Development as Experiment: Science and State Building in Late Colonial and Postcolonial Africa, 1930–1970"; and Vinh-Kim Nguyen, "Government-by-Exception."

5. The World Fertility Survey, which began in 1972, organized by USAID and the UN surveyed over 350,000 women in 62 countries. On the global facts of life see Stacy Leigh Pigg, "Globalizing the Facts of Life."

6. On free clinics, and community health centers, see Jennifer Nelson, "'Hold Your Head up and Stick Our Your Chin': Community Health and Women's Health in Mound Bayou, Mississippi"; Naomi Rogers, "'Caution: The Ama May Be Dangerous to Your Health': The Student Health Organizations (Sho) and American Medicine, 1965–1970." The term *medical-industrial complex* originates from the HealthPAC publication by Barbara Ehrenreich and John Ehrenreich, *The American Health Empire: Power, Profit and Politics*. Barbara Ehrenreich went on to publish another work which had a major influence in how women health activists in the 1970s narrated the history of medicine.

7. See, for example, Jennifer Nelson, *Women of Color and the Reproductive Rights Movement*.

8. See Phil Brown and Edwin Mikkelsen, *No Safe Place: Toxic Waste, Leukemia, and Community Action*; Michelle Murphy, *Sick Building Syndrome and the Problem of Uncertainty: Environmental Politics, Technoscience, and Women Workers*.

9. Shulamith Firestone, *The Dialectic of Sex: The Case for Feminist Revolution*.

10. See, for example, Sonia Alvarez, "Latin American Feminisms Go Global: Trends of the 1990s, Challenges for the New Millennium"; and Millie Thayer, "Transnational Feminism: Reading Joan Scott in the Brazilian Sertao."

11. The declaration is reprinted in FINRRAGE-UBINIG International Conference 1989, "Declaration of Comilla."

12. Farida Akhter, *Depopulating Bangladesh: Essays on the Politics of Fertility.*

13. FINRRAGE-UBINIG International Conference 1989, "Declaration of Comilla."

14. Personal communication, 2006.

15. I discuss Sanger's version of control in greater depth in Michelle Murphy, "Liberation through Control in the Body Politics of U.S. Radical Feminism."

16. *Reproductive health* as a term emerges in the published record in the early 1970s within the medical literature, particularly in works dealing with questions of family-planning services. The 1977 presidential address of the American Society of Obstetricians and Gynecologists called for the designation of "reproductive health" as a specialty of the field that recognized their "major responsibility" to help "limit the numbers of children being born." C. L. Randall, "The Obstetrician-Gynecologist and Reproductive Health." For some of the background to the history of the feminist concept of "reproductive health" see both chapter 3 in this book, and Saul Halfon, *The Cairo Consensus: Demographic Surveys, Women's Empowerment and Regime Change in Population Policy*; Betsy Hartmann, *Reproductive Rights and Reproductive Wrongs: The Global Politics of Population Control*; and Rosalind Petchesky, *Global Prescriptions: Gender, Health and Human Rights.*

17. The term *generation* preceded the term *reproduction*. The term was first developed by Buffon, and also notably developed by Hegel. Linnaeus, though he used the term *generation*, importantly entangled his work with notions of nature as marketplace. On reproduction's eighteenth-century origins, see Ludmilla Jordanova, "Interrogating the Concept of Reproduction in the Eighteenth Century"; Staffan Mueller-Wille and Hans-Jörg Rheinberger, *Heredity Produced: At the Crossroads of Biology, Politics, and Culture, 1500–1870*; and Tilottama Rajan, "Dis-Figuring Reproduction: Natural History, Community, and the 1790s Novel."

18. Londa Schiebinger, "Why Mammals Are Called Mammals: Gender Politics in Eighteenth Century Natural History."

19. Staffan Mueller-Wille, "Nature as a Marketplace: The Political Economy of Linnaean Botany"; Margaret Schabas, "Adam Smith's Debts to Nature"; Margaret Schabas, *The Natural Origins of Economics.*

20. *Biopolitics* as a term is developed by Foucault in his *The Birth of Biopolitics: Lectures at the College De France, 1978–79*; *History of Sexuality*, vol. 1: *An Introduction*; *Security, Territory, Population: Lectures at the College De France, 1977–1978*; and *Society Must Be Defended: Lectures at the Colleges De France, 1975–76.*

21. In emphasizing multiple genealogies and multiple biopolitical modes, I'm drawing here on a feminist historiography that highlights the need for multiple periodizations and genealogies, currents in transnational feminist studies that call for attention to "discrepant dislocations" and "scattered hegemonies," insights within technoscience studies into the multiplicity of becoming, as well as theoretical work of Gilles Deleuze and his collaborators. Elsa Barkley Brown, "Polyrhythms and Improvisation: Lessons for Women's History"; Deleuze and Félix Guattari, *A Thousand Plateaus: Capitalism and Schizophrenia*; Gilles Deleuze and Claire Parnet, *Dialogues*;

Inderpal Grewal and Caren Kaplan, *Scattered Hegemonies: Postmodernity and Transnational Feminist Practices*; Mary John, *Discrepant Dislocations: Feminism, Theory and Postcolonial Histories*; Annemarie Mol and John Law, *Complexities: Social Studies of Knowledge Practices*; Murphy, *Sick Building Syndrome and the Problem of Uncertainty: Environmental Politics, Technoscience, and Women Workers*.

22. Yet, it is precisely the fields of feminist and colonial/postcolonial studies that Foucault's work has been central to. For examples see Carolyn Ramazanoglu, *Up against Foucault: Explorations of Some Tensions between Foucault and Feminism*; David Scott, "Colonial Governmentality"; Ann Laura Stoler, *Carnal Knowledge and Imperial Power: Race and the Intimate in Colonial Rule*; and Stoler, *Race and the Education of Desire: Foucault's* History of Sexuality *and the Colonial Order of Things*.

23. Cindi Katz, a geographer, builds on Donna Haraway's theorization of situated knowledge by bringing in concerns from her field of critical geography to elaborate something she calls "counter topographies" as a description of her own knowledge-making project. Cindi Katz, *Growing up Global: Economic Restructuring and Children's Everyday Lives*; Katz, "On the Grounds of Globalization: A Topography for Feminist Political Engagement." Katz's notion of counter-topographies, therefore, notably differs from Haraway's important theorization of situated knowledge in that for Katz it describes her own analytic tactics as a geographer, while for Haraway all knowledge making is situated knowledge even if it does not mark itself as such. Donna Haraway, "Situated Knowledges: The Science Question in Feminism and the Privilege of Partial Perspective." The sense of biopolitical topology as recursive origami developed in this book, like Haraway, similarly seeks to describe a particular way of both analyzing and participating in biopolitics, but also seeks to name the complex struggles extensive in time and space that constitute past biopolitical pluralities, whether or not those formations marked themselves as political.

24. Drawing on the work of critical geographers, I have chosen topology over other terminological possibilities, such as networks (as used in actor-network-theory), or situated (following Donna Haraway's development of situated knowledge), or topography. For related efforts to use topology to describe spatialized multidimensional transformations, see Oliver Belcher et al., "Everywhere and Nowhere: The Exception and the Topological Challenge to Geography"; Steven Collier, "Topologies of Power: Foucault's Analysis of Political Government beyond 'Governmentality'"; Steve Hinchliffe, "Scalography and Worldly Assemblies"; Brian Massumi, *Parables for the Virtual: Movement, Affect, Sensation*; and Mol and Law, *Complexities: Social Studies of Knowledge Practices*.

25. Rendering the full repleteness of biopolitics is impossible, so I prefer to imagine my approach as a more humble method of origami, or better yet a recursive origami, conducted as a kind of handmade craft work, accomplished through the limits of paper, that seeks to follow folds within folds, while at the same time unavoidably refolding. In other words, what I am positing as a biopolitical topology is rendered legible here only partially through the more modest acts of analysis that

incompletely engage a multidimensional landscape of layered entanglements and historically drawn borders. Here, I am inspired not by the precise, rule-bound techniques of conventional origami, but the textured shapes produced by what practitioners called "crumpled origami." In turn, the methodology I am calling recursive origami does not begin with a smooth flat plane, but always enters the archive of the past as already refolded, always crumpled, unavoidably uneven — the world already complexly articulated through race, sex, capital, and nation. Analytical acts of unfolding chart how practices and phenomena (such as Pap smears and cervical cancer) are already made up of many other folds stretching far beyond the object itself. In this spirit the chapters investigate the multiple material, political, and economic conditions that are pressed into an objects, subjects, and practices.

When a fold is pulled open, an origami figure twists into a different form. Yet the creases set by the previous folds remain, marking how different formations coexist in the multidimensional shape. Each unfolding underlines that phenomena are not just multiply made, but are many things at once within layered and convoluted histories. Each fold makes a relationship. It binds two sides along a crease; it articulates proximities; it appropriates past folds into new arrangements. Folds divide and layer in complex spatialized relationships. Engaging a topological past by reimagining historical work as a kind of origami, then, impinges on the ontological stakes of historical writing itself: how historical narratives also participate in remaking the phenomena they engage. Thus, recursive origami is not the contrary of folding (that is unmaking or undoing), but rather is meant as a critical, yet, modest, analysis that maps how an element or politics that seems necessary and self-evident in one configuration becomes bendable, pliable, and questionable in another configuration. Is it possible to make another set of folds and build a different configuration all together?

26. On necropolitics and precariousness, see Giorgio Agamben, *Homo Sacer: Sovereign Power and Bare Life*; Judith Butler, *Precarious Life: The Powers of Mourning and Violence*; and Achille Mbembe, "Necropolitics."

27. Here I am building on Judith Butler's insight that what gets fixed as matter, as outside of social construction, is an effect of power, which she calls "materialization." Judith Butler, *Bodies That Matter: On the Discursive Limits of "Sex"*.

28. Joan Scott, *Only Paradoxes to Offer: French Feminists and the Rights of Man*.

29. A corollary claim is that subject positions were carved out of "biological" kinds, such as raced, sexed, gendered subjects, as well as subject positions articulated through an evolutionary scheme such as primitive and civilized, or out of degenerate schema, such as criminal and insane, and so on.

30. My approach here builds on and yet veers substantially from much of the historiography on reproductive politics, which more often focuses on questions of law and rights or, within the history of technoscience, on techniques and epistemologies. My approach here is also similarly related to Marxist feminist work that focuses on the devaluation of reproduction in capital formations, which has tended to univocally see reproduction as an unvalued, free, or devalued form of labor that capital

rests on and hence expropriates. Building on such work I am interested in asking how reproduction is unevenly valued in modes both explicitly economic, as well as extra-economic. Thus, I assume here that capital formations do rest on reproduction, not because it is simply devalued, but because reproduction is a complex and extensive process generative of value, entangled with multiple genealogies of value-making; moreover, the domain of reproduction is produced as such in the tangle of those relations.

31. "Composite Report of the President's Committee to Study the United States Military Assistance Program."

32. "Making Foreign Aid Work," *New York Times*, July 26, 1959. E8.

33. Akhter, *Depopulating Bangladesh: Essays on the Politics of Fertility*; Kamran Asdar Ali, *Planning the Family in Egypt: New Bodies, New Selves*; Ruhul Amin et al., "Family Planning in Bangladesh, 1969–1983"; Susan Greenhalgh and Edwin Winckler, *Governing China's Population: From Leninist to Neoliberal Biopolitics*; Halfon, *The Cairo Consensus*; Betsy Hartmann and Hilary Standing, *Food, Saris and Sterilization: Population Control in Bangladesh*; UBINIG, *Violence of Population Control*.

34. Franz Fanon, "Medicine and Colonialism."

35. As is discussed at greater length in chapter 4, USAID family-planning assistance, which tended to go to nonprofit family-planning organizations such as International Planned Parenthood, or other organizations that today are called NGOs, was a site in which neoliberal formations crystallized early.

36. On militarism's extension into seemingly nonmilitarized realms of daily life, see Cynthia Enloe, *Maneuvers: The International Politics of Militarizing Women's Lives*.

37. For a more extended argument see Michelle Murphy, "The Economization of Life."

38. While Mitchell makes the suggestion earlier, Bergeron provides a deeper argument and analysis. Suzanne Bergeron, *Fragments of Development: Nation, Gender, and the Space of Modernity*; Timothy Mitchell, "Fixing the Economy."

39. Daniel Speich, "The World of GDP: Historicizing the Epistemic Space of Postcolonial Development."

40. Lyndon B. Johnson, "Address in San Francisco at the 20th Anniversary Commemorative Session of the United Nations" (1965). On Johnson's speech, see Rickie Solinger, *Wake up Little Susie: Single Pregnancy and Race before Roe v. Wade*. Johnson's statement is similar to an earlier calculus offered by Dr. Joseph Beasley, whose family-planning projects in Louisiana administered by the Ford Foundation and paid for with state, and later federal, funds were touted as a national and international model. Beasley's studies claimed that for every dollar spent on family planning, the state saved over thirteen dollars on welfare costs. For an excellent account of federal family-planning policy, as well as Beasley's work, see Donald Crichtlow, *Intended Consequences: Birth Control, Abortion, and the Federal Government in Modern America*; and Martha Ward, *Poor Women, Powerful Men: American's Great Experiment in Family Planning*.

41. See Lawrence Summers, "The Most Influential Investment," 108.

42. On the history of demographics entanglements with eugenics, see Ramsden, "Carving up Population Science: Eugenics, Demography and the Controversy over the 'Biological Law' of Population Growth"; Edmund Ramsden, "Social Demography and Eugenics in the Interwar United States"; and Simon Sretzer, "The Idea of Demographic Transition and the Study of Fertility: A Critical Intellectual History."

43. See, for example, the Moynihan Report.

44. Donald Crichtlow, *Intended Consequences* and Ruth Feldstein, *Motherhood in Black and White: Race and Sex in American Liberalism, 1930–1965.*

45. Alexandra Minna Stern, "Sterilized in the Name of Public Health: Race, Immigration, and Reproductive Control in Modern California"; and Elena Gutierrez, *Fertile Matters: The Politics of Mexican-Origin Women's Reproduction.*

46. On the "culture of poverty" see Ruth Feldstein, *Motherhood in Black and White*; and Alice O'Connor, *Poverty Knowledge: Social Science, Social Policy, and the Poor in Twentieth-Century U.S. History.*

47. Adele Clarke et al., "Biomedicalization: Technoscientific Transformations of Health, Illness, and U.S. Biomedicine." In this book, I'm temporally extending the process of biomedicalization earlier, back to the 1970s, whereas Clarke and her collaborators have dated the crystallization of biomedicalization as a dominant mode to the mid-1980s. While I largely agree with their time frame, in this book I am interested in looking at the emergence of biomedicalization, rather than its instantiation as the dominant logic.

48. Catherine Waldby, "Stem Cells, Tissue Cultures and the Production of Biovalue."

49. Sarah Franklin, *Dolly Mixtures: The Remaking of Genealogy.*

50. Ibid.; Charis Thompson, *Making Parents: The Ontological Choreography of Reproductive Technologies*; Waldby, "Stem Cells, Tissue Cultures and the Production of Biovalue."

51. See also Michelle Murphy, "Sex and Gender."

52. For example, see Vin-Kim Nguyen, "Antiretroviral Globalism, Biopolitics, and Therapeutic Citizenship"; Adriana Petryna, Andrew Lakoff, and Arthur Kleinman, *Global Pharmaceuticals: Ethics, Markets, Practices*; Kaushik Sunder Rajan, *Biocapital: The Constitution of Postgenomic Life.*

53. I make this argument in chapter 4.

54. Partha Chatterjee, *The Politics of the Governed: Reflections on Popular Politics in Most of the World.*

55. Adriana Petryna, *Life Exposed: Biological Citizens after Chernobyl.*

56. For a quick genealogy of the term, see Emilio Mordini, "Biowarfare as a Biopolitical Icon."

57. W. E. B. Du Bois, *The Souls of Black Folks.*

58. Gayatri Chakravorty Spivak, "In a Word: Interview."

Chapter 1. Assembling Protocol Feminism

1. Montreal Health Press, *The Birth Control Handbook*. List from West Coast Sisters, "How to Start Your Self-Help Clinic, Level II."

2. Frances Hornstein, Carol Downer, and Shelly Farber, *Self Help and Health: A Report*.

3. On protocols in biomedicine, see Marc Berg and Stefan Timmermans, "Standardization in Action: Achieving Local Universality through Medical Protocols." On protocols within information technology as a decentered logic of governmentality, see Alexander Galloway, *Protocol: How Control Exists after Decentralization*.

4. Lorraine Rothman, interview, October 23, 1999.

5. Carol Downer, interview, October 24, 1999.

6. Medical supply stores were quite willing to sell their merchandise, and thus self help clinics met with little resistance in that regard. An offshoot of these tests was that some of the women who tested positive asked for help attaining abortions. Downer and Rothman then negotiated with a local hospital to provide abortions under strict conditions—specifying from what the doctor should wear to the technique used and, importantly, the price. In return, the self help clinic received fifteen dollars, provided the initial counseling, and accompanied the woman through the whole procedure. Downer, interview, October 24, 1999.

7. The first self help group in Los Angeles initially met with the idea of forming an underground abortion service modeled on Jane (discussed later in the chapter), and inspired by the Californian group called the "Army of Three"—Patricia Maginnis, Rowena Gurner, and Lana Clarke Phelan, who practiced a confrontational, satirical feminist stand on abortion in the 1960s, in the years before the explosion of radical feminism—offered inspiration. Their tactics included offering "abortion classes" and publishing the *Abortion Handbook for Responsible Women*, written both to outrage and inform, which advised women on the technicalities of how to find a "back yard" abortionist, fake a hemorrhage, or induce an abortion with one's fingers. Patricia Maginnis and Lena Clarke Phelan, *The Abortion Handbook for Responsible Women*. Downer had been mentored by this group, and had gone on to research and tour abortion services offered both legally and illegally on the West Coast. Lana Clarke Phelan was chair of the LA NOW chapter's taskforce on abortion when Downer first met her. Pat Maginnis's original group was called the Society for Humane Abortion. Maginnis and Gurner together founded the Association for the Repeal of Abortion Laws, which grew into the National Association for the Repeal of Abortion Laws and then after *Roe v. Wade*, changed to the less radical National Abortion Rights Action League. On the Army of Three, see Ninia Baehr, *Abortion without Apologies: A Radical History for the 1990s*.

8. For formal protocols on feminist abortion, see Feminist Women's Health Center, *Abortion in a Clinical Setting*. To meet increasingly strict state licensing requirements,

clinics of the Federation of Feminist Health Centers (discussed later in the chapter) developed "the black book" as its official account of technical protocols.

9. In its claims to a "revolutionary" politics that would "seize" their own bodies, the feminist self help movement was exemplary of much radical feminisms of the late 1960s and 1970s. *Radical* as a term was used in the late 1960s to describe a political subject position associated with "extreme" social change that goes to the "root." *Radical feminism* tended to be a self-nominated term used by feminists who stressed patriarchy as an origin of oppression, and was most often embraced as a moniker by women for whom sexism, rather than questions of race or class, came to the fore. However, as a wider term, it designates a heterogeneous thread of feminism of the late 1960s and 1970s that set itself apart from liberal and strictly Marxist feminisms, and was sometimes deployed by feminisms organized as an element of a raced or ethnic identity or community. Radical feminism was extremely heterogeneous, tending to be practiced in small, local, independently formed cells. Ideologically, however, radical feminists, tended to believe that women were universally, even if multiply and unevenly, oppressed; that the ultimate root cause of this oppression was patriarchy, not just capitalism; and that the solution was tearing down patriarchal social structures, not reform.

While historians of radical feminism have tended to concentrate on the writings penned by college-educated white women in the Northeast in the 1960s, more diverse constellations of radical feminists scattered over the United States and Canada also founded many issue-oriented projects—rape crisis centers, battered women's shelters, feminist bookstores, and feminist health clinics—which set out to create alternative women-controlled institutions. Moreover, unlike the closed vanguardism of many of the early cells, in the 1960s, these projects typically set out to appeal widely to women, providing services for diverse constituencies of women who did not necessarily see themselves as feminists. Many of these services, such as feminist health clinics, attempted at their inception to prefigure within themselves the kinds of social relations they wished for. Thus, this formulation of service-oriented radical feminism was more diverse and often involved politicizing protocol. Alice Echols in her important history of radical feminism portrays these projects as a form of deradicalized cultural feminism that followed after radical feminism. I think that this is too rigid a designation that overlooks a continuity with efforts to build feminist counterinstitutions within consciousness-raising groups of early radical feminist cells. Echols is particularly critical of the Feminist Economic Network (FEN). Laura Brown, director of the Oakland Feminist Women's Health Center, was a prominent advocate of feminist businesses and, with the Detroit Feminist Federal Credit Union, formed the controversial FEN. The FEN, however, is far from representative of the diversity of feminist projects that were a part of radical feminism, most of which were nonprofits. Alice Echols, *Daring to Be Bad: Radical Feminism in America.*

On the reticulate, decentralized, and segmented form of late twentieth-century social movements, see Luther P. Gerlach and Virginia H. Hine, *People, Power, Change: Movements of Social Transformation.*

10. This point could also be broadened to include the proliferation of sexed counter-conduct. The term *counter-conduct* here is used in relation to Foucault's notion of governmentality as techniques for the "conduct of conduct," which can be rephrased as the directing of practice, which can further be redefined as protocols. Important in Foucault's notion of governmentality is its disassociation from the state as the site of governance and his insistence on a more general dispersal of such techniques in many domains of life. As Arnold Davidson shows, Foucault's notion of the "conduct of conduct" was accompanied by his interest in and calls for "counter-conduct," practices that attempt to inhabit and interrupt in ways antagonistic to hegemonic conduct. Arnold Davidson, "In Praise of Counter-Conduct."

11. The sociologists Marc Berg and Stefan Timmerman define medical protocol as a "technoscientific script that crystallizes multiple trajectories." They draw attention to how protocols organize the multiple trajectories (past and present) that meet in the standardized use of a technoscientific artifact within biomedicine. Berg and Timmermans, "Standardization in Action."

12. Gilles Deleuze and Felix Guattari, *A Thousand Plateaus: Capitalism and Schizophrenia*, 99.

13. On this final example, see Halfon, *The Cairo Consensus*, as well as, chapters 3 and 4 of this book.

14. See, for example, Kathy Davis, *The Making of* Our Bodies, Ourselves: *How Feminism Travels across Borders*; and Wendy Kline, *Bodies of Knowledge: Sexuality, Reproduction, and Women's Health in the Second Wave*. Also, Judith Houck is undertaking a social history of feminist health clinics in the United States.

15. The concept of biopolitical topologies is treated in the introduction.

16. "Taking back turf" was a phrase that circulated amongst activists, rather than in materials for public consumption. Dido Hasper, Shauna Heckert, and Eilleen Schnitger, Chico Feminist Women's Health Center, group interview, November 1999.

17. On the marking of itself as "outside" see chapter 4.

18. Margaret Sanger, "The Birth Control League"; Ellen Chesler, *Woman of Valor: Margaret Sanger and the Birth Control Movement in America*.

19. Claudia Dreifus, "Introduction," in *Seizing Our Bodies*, edited by Claudia Dreifus (New York: Vintage, 1977), xxxi. For a genealogy of this credo, see Murphy, "Liberation through Control in the Body Politics of U.S. Radical Feminism."

20. Santa Cruz Women's Health Collective, *The Self Help Booklet* (n.d., ca. early 1970s), Boston Women's Health Collective Archive, file "Self-Help: Gynecological."

21. Robin Morgan, introduction to *Circle One*, 3.

22. Donna Haraway, *Modest-Witness@Second-Millennium*, 193.

23. On debates over reproductive rights in transnational feminist circuits, see Farida Akhter, "Reproductive Rights: A Critique from the Realities of Bangladeshi

Women"; Sonia Correa, *Population and Reproductive Rights: Feminist Perspectives from the South*; Betsy Hartmann, *Reproductive Rights and Reproductive Wrongs*; Petchesky, *Global Prescriptions: Gender Health and Human Rights*; and Charlotte Rutherford, "Reproductive Freedoms and African American Women."

24. Beverly Smith, "Black Women's Health: Notes for a Course."

25. Lawrence Cohen, "Operability, Bioavailability, and Exception."

26. I see a parallel between Foucault's description of governmentality as the "conduct of conduct" (conduire des conduits) and my attention here to historicizing biopolitics as the "politics of politics." Foucault, "Le sujet et le pouvoir," and "The Subject and Power"; Colin Gordon, "Government Rationality: An Introduction."

27. Foucault was not the first to use this term. For a brief genealogy of the term as it circulated contemporaneously to Foucault and briefly preceded his use, see Kolson Schlosser, "Bio-Political Geographies."

28. Foucault, *History of Sexuality*, vol. 1.

29. Ibid.

30. The notion of technopolitics comes from Gabrielle Hecht, *The Radiance of France: Nuclear Power and National Identity after World War II.*

31. Here I am building on large critical literature concerning whiteness as historically and geographically specific formations. See Richard Delgado and Jean Stefancic, *Critical White Studies: Looking Beyond the Mirror*; Richard Dyer, "White"; George Lipsitz, *The Possessive Investment in Whiteness: How White People Profit from Identity Politics*; and Martha Mahoney, "Residential Segregation and White Privilege." A much smaller literature on whiteness and science includes Warwick Anderson, *The Cultivation of Whiteness: Science, Health and Racial Destiny in Australia*; Lisa Bloom, "Constructing Whiteness: *Popular Science* and *National Geographic* in the Age of Multiculturalism"; and Murphy, "Uncertain Exposures and the Privilege of Imperception: Activist Scientists and Race at the U.S. Environmental Protection Agency" (*Osiris* 19 [2004]).

32. The term *possessive individualism* comes from the theorist C. B. McPherson, *The Political Theory of Possessive Individualism: Hobbes to Locke*.

33. Ronald Formisano, *Boston against Busing: Race, Class, and Ethnicity in the 1960s and 1970s.*

34. Boston Women's Health Book Collective, *Our Bodies, Ourselves: A Book by and for Women*, 2.

35. On the Boston Women's Health Collective and the Somerville Health Center, see Davis, *The Making of Our Bodies, Ourselves*; and Susan Reverby, "Alive and Well in Somerville, Mass."

36. Combahee River Collective, "Combahee River Collective Statement"; also reprinted as "A Black Feminist Statement" in Gloria Hull, Patricia Bell-Scott, and Smith Barbara, *All the Women Are White, All the Blacks Are Men, but Some of Us Are Brave: Black Women's Studies*.

37. On disidentification and queer women of color critique, see Roderick Ferguson, *Aberrations in Black: Toward a Queer of Color Critique*; and José Esteban Muñoz, *Disidentifications: Queers of Color and the Performance of Politics.*

38. Combahee River Collective, "Combahee River Collective Statement."

39. Beverly Smith, "Black Women's Health."

40. Beverly Smith, "The Development of an Ongoing Data System at a Women's Health Center."

41. "Women's Community Health Center, Statement" (n.d., ca. mid-1970s), Boston Women's Health Collective Archive, file "Self Help: Gynecology."

42. Theodor Nelson, *Computer Lib: You Can and Must Understand Computers Now / Dream Machines: New Freedoms through Computer Screens—A Minority Report.*

43. Federation of Feminist Women's Health Centers, *A New View of a Woman's Body*, 17.

44. On the fashioning of explicitly white feminisms, see Louise Michele Newman, *White Women's Rights: The Racial Origins of Feminism in the United States.*

45. Howard Winant, "Behind Blue Eyes: Whiteness and Contemporary U.S. Racial Politics.

46. Doug Rossinow, *The Politics of Authenticity: Liberalism, Christianity and the New Left in America.*

47. My analysis of the Combahee River Collective is deeply indebted to critical insights of Grace Kyungwon Hong, *The Ruptures of American Capital: Women of Color Feminism and the Culture of Immigrant Labor*. See also Duchess Harris, "From the Kennedy Commission to the Combahee Collective: Black Feminist Organizing, 1960–80"; and Brian Norman, "'We' in Redux: The Combahee River Collective's 'A Black Feminist Statement.'"

48. See, for example, African American Women for Reproductive Freedom, "We Remember." On the Moynihan Report (1965) and social science valuations of black motherhood, see Feldstein, *Motherhood in Black and White*; and Ferguson, *Aberrations in Black.*

49. This phrase is from Hammer's speech at the Democratic National Convention of 1964. For uses of the phrase, see Byllye Avery, "A Question of Survival / A Conspiracy of Silence: Abortion and Women's Health"; Angela Davis, "Sick and Tired of Being Sick and Tired: The Politics of the National Black Women's Health Project"; and Susan Smith, *Sick and Tired of Being Sick and Tired: Black Women's Health Activism in America.*

50. This is a phrase taken from the work of poet, feminist, and peace activist Barbara Deming. Deming, *We Cannot Live without Our Lives*. A photo of the protest and banner was used as the cover for the issue of *Radical America*, where the pamphlet was republished. Combahee River Collective, "Why Did They Die?" On survival in black, queer, feminist politics and poetics, see Alexis Pauline Gumbs, "We Can Learn to Mother Ourselves: The Queer Survival of Black Feminism 1968–1996."

51. Agamben, *Homo Sacer*; Dorothy Roberts, *Killing the Black Body: Race, Reproduc-*

tion, and the Meaning of Liberty; Sarah Lochlann Jain, *Injury: The Politics of Product Design and Safety Law in the United States*; Andrea Smith, *Conquest: Sexual Violence and American Indian Genocide*; Mbembe, "Necropolitics"; Melissa Wright, *Disposable Women and Other Myths of Global Capitalism*.

52. Agamben, *Homo Sacer*; Avery, "Breathing Life into Ourselves"; Avery, "A Question of Survival / A Conspiracy of Silence: Abortion and Women's Health"; Avery, "Who Does the Work of Public Health"; Jain, *Injury*; Mbembe, "Necropolitics"; Wright, *Disposable Women and Other Myths of Global Capitalism*.

53. Deborah Gray White, *Too Heavy a Load: Black Women in Defense of Themselves, 1894–1994*, 225.

54. Patricia Harden et al., "The Sisters Reply," 2.

55. For pivotal works that theorized this divide between the liberal subject of much white feminist theorizing, and the contradictory subject—what Chela Sandoval calls differential consciousness—see Gloria Anzaldúa, *Borderlands / La Frontera: The New Mestiza*; Cherrie Moraga and Gloria Anzaldúa, *This Bridge Called My Back: Writings by Radical Women of Color*; and Chela Sandoval, *Methodology of the Oppressed*.

56. Helen Rodriguez-Trias, "The Women's Health Movement: Women Take Power."

57. My reconstruction of these events is aided by interviews with Lorraine Rothman and Carol Downer, October 23 and October 24, 1999.

58. Suzanne Gage, an important member of the early Los Angeles feminist self help, who did most of the illustrations for publications and later helped establish feminist lesbian health services, had previous interactions with Jane as a university student from Illinois. See note 7 above for Downer's ties to the Army of Three. Rothman had been involved in the National Organization of Women in Orange County.

59. Lorraine Rothman, interview, October 23, 1999.

60. Brian Beaton, personal communication, May 2007.

61. There was, and still is, an absence of reflection as to how Los Angeles feminist self help of the 1970s functioned through whiteness and its hegemonic normality. This lack of reflexivity about racism is pervasive in white feminist memoirs of this era. Barbara Smith, "'Feisty Characters' and 'Other People's Causes': Memories of White Racism and U.S. Feminism."

62. They also differentiated their hands-on approach from the public education work of the Boston Women's Health Book Collective.

63. Downer, interview, October 24, 1999.

64. Ibid.

65. Dido Hasper, Shauna Heckert, and Eilleen Schnitger, Chico Feminist Women's Health Center, group interview, November 1999.

66. This graphic was given to Downer when she was acquitted of charges of practicing medicine without a license. For an analysis of the work of Wonder Woman in this image, see Haraway, *Modest Witness*, 194–96.

67. While the notion of "self help" in the United States dates at least to late nineteenth-century popularity of Samuel Smiles' book, *Self Help* (1882), which offered

the maxim "heaven helps those who help themselves" as a call for even the most "humble" to engage in bootstrapping "energetic action" to improve themselves. Historians of medicine, immigration, gender, and African American life in the United States have excavated a large history of nineteenth-century mutual assistance clubs, mutual insurance associations, racial mutual aid, and other such practices crafted by working-class and rural communities, in which women's labor was often a cornerstone.

68. Here I am building on the insights offered in Monica Greco, "The Politics of Indeterminacy and the Right to Health."

69. Ann Laura Stoler, "Racial Histories and Their Regimes of Truth."

70. This term was first used in John Ehrenreich and Barbara Ehrenreich, *The American Health Empire: Power, Profits, and Politics.*

71. Thomas Schultz and Gary Becker developed the notion of human capital in this period. See also Murphy, "The Economization of Life."

72. On virtual pathology, see Kathryn Morgan, "Contested Bodies, Contested Knowledges: Women, Health and Promotion." On the rise of the at-risk patient, see Nikolas Rose, *The Politics of Life Itself: Biomedicine, Power, and Subjectivity in the Twenty-First Century.* On prescribing by numbers, see Jeremy Greene, *Prescribing by Numbers: Drugs and the Definition of Disease.*

73. Clarke et al. dates the crystallization of biomedicalization to the mid-1980s. While I agree with this periodization, many of the components making up this crystallization were emergent in the 1960s and '70s. Thus, I extend the use of the term to describe the earlier inauguration of attributes of biomedicalization. Clarke et al., "Biomedicalization."

74. On the "racial state," see chapter 3 of this book, and David Theo Goldberg, *The Racial State.*

75. On healthism, see R. Crawford, "Healthism and the Medicalization of Everyday Life"; and Greco, "The Politics of Indeterminacy and the Right to Health." I use the term *stratified biomedicine* to help capture the sense of the unevenness of biomedicine.

76. Marc Berg, "Turning a Practice into a Science: Reconceptualizing Postwar Medical Practice."

77. Donald Crichtlow, *Intended Consequences: Birth Control, Abortion, and the Federal Government in Modern America.*

78. Matthew Connelly, *Fatal Misconception: The Struggle to Control World Population*; Hartmann, *Reproductive Rights and Reproductive Wrongs*; Rickie Solinger, *Beggars and Choosers: How the Politics of Choice Shapes Adoption, Abortion, and Welfare in the United States.*

79. On entanglements between birth control, modernization theory, and development see Kamran Asdar Ali, *Planning the Family in Egypt: New Bodies, New Selves*; Nilanjana Chatterjee and Nancy Riley, "Planning an Indian Modernity: The Gendered Politics of Fertility Control"; and Omnia El Shakry, *The Great Social Laboratory: Subject of Knowledge in Colonial and Postcolonial Egypt.*

80. Crichtlow, *Intended Consequences*; Johanna Schoen, *Choice and Coercion: Birth Control, Sterilization, and Abortion in Public Health and Welfare*.

81. Gerald Bernstein, "The Los Angeles Experience"; Lynn Landman, "Los Angeles' Experiment with Functional Coordination: A Progress Report."

82. Joy Dryfoos, "Planning Family Planning in Eight Metropolitan Areas."

83. This does not include independent unassociated feminist health centers. Bernstein, "The Los Angeles Experience"; Landman, "Los Angeles' Experiment with Functional Coordination."

84. Randall Hulbert and Robert Settlage, "Birth Control and the Private Physician."

85. Morton Silver, "Birth Control and the Private Physician."

86. Elena Gutierrez, "Policing 'Pregnancy Pilgrims': Situating the Sterilization Abuse of Mexican-Origin Women in Los Angeles County"; Alexandra Minna Stern, "Sterilized in the Name of Public Health: Race, Immigration, and Reproductive Control in Modern California."

87. Alondra Nelson, *Body and Soul: The Black Panther Party and the Fights against Medical Discrimination*; Peter Rudd, "The United Farm Workers Clinic in Delano, Calif.: A Study of the Rural Poor"; Barbara Russell and Lynn Lofstrom, "Health Clinic for the Alienated."

88. Marie Branch and Phyllis Paxton, *Providing Safe Nursing Care for Ethnic People of Color*; Naomi Rogers, "'Caution: The Ama May Be Dangerous to Your Health.'"

89. See, for example, Jennifer Nelson, "'Hold You Head up and Stick Our Your Chin': Community Health and Women's Health in Mound Bayou, Mississippi."

90. Franz Fanon, "Medicine and Colonialism"; Che Guevara, "On Revolutionary Medicine"; Rogers, "'Caution'."

91. Governor's Commission on the Los Angeles Riots, "Violence in the City—An End or a Beginning?"; "Watts Riots Anniversary: Community Fights to Save Jobs, Watts Health Foundation Created in Response to Violence."

92. Perlita Dicochea, "Chicana Critical Rhetoric: Recrafting La Causa in Chicana Movement Discourse, 1970–1979"; Sherna Berger Gluck et al., "Whose Feminism, Whose History? Reflections on Excavating the History of (the) US Women's Movement"; Nancy Matthews, *Confronting Rape: The Feminist Anti-Rape Movement and the State*; Laura Pulido, *Black, Brown, Yellow and Left: Radical Activism in Los Angeles*; Leslie Reagan, "Crossing the Border for Abortions: California Activists, Mexican Clinics, and the Creation of a Feminist Health Agency in the 1960s."

93. Pulido, *Black, Brown, Yellow and Left*.

94. Carol Downer, "Self Help: What Is It?"

95. Paul Rabinow, "Artificiality and Enlightenment: From Sociobiology to Biosociality."

96. See chapter 2, and Francesca Polletta, *Freedom Is an Endless Meeting*.

97. Sara Evans, *Personal Politics*; Charles Payne, *I've Got the Light of Freedom: The Organizing Tradition and the Mississippi Freedom Struggle*; Daniel Perlstein, "Teaching Freedom: SNCC and the Creation of the Mississippi Freedom Schools."

98. Steve Biko, "The Definition of Black Consciousness"; Jeffrey Ogbar, *Black Power: Radical Politics and African American Identity*; Barney Pityana, *Bounds of Possibility: The Legacy of Steve Biko and Black Consciousness*; Barbara Ransby, *Ella Baker and the Black Freedom Movement: A Radical Democratic Vision*; Mary Rolinson, *Grassroots Garveyism: The Universal Negro Improvement Association in the Rural South*; Stephen Ward, "The Third World Women's Alliance: Black Feminist Radicalism and Black Power Politics."

99. Franz Fanon, *Black Skin, White Masks*; Albert Memmi, *The Colonizer and the Colonized*.

100. Citations of Mao among American radicals typically were filtered through William Hinton, *Fanshen: A Documentary of Revolution in a Chinese Village*.

101. Kathie Sarachild, "Consciousness-Raising: A Radical Weapon."

102. Rossinow. *The Politics of Authenticity*.

103. Ellen Herman, *The Romance of American Psychology: Political Culture in the Age of Experts*.

104. I draw on this work to frame this section. Laura Kim Lee, "Changing Selves, Changing Society: Human Relations Experts and the Invention of T Groups, Sensitivity Training, and Encounter in the United States, 1938–1980."

105. Kurt Lewin and Ronald Lippitt, "An Experimental Approach to the Study of Autocracy."

106. Examples here range from bestsellers to Theodore Adorno's work on the F-test. Theodore Adorno et al., *The Authoritarian Personality*; William Whyte, *The Organizational Man*.

107. Lee, "Changing Selves, Changing Society."

108. See, in particular, Kenneth Benne, *A Conception of Authority*; and Benne, "Democratic Ethics in Social Engineering."

109. Richard Burke and Warren Bennis, "Changes in Perception of Self and Others during Human Relations Training," 166.

110. Benne, "History of the T Group in the Laboratory Setting," 82.

111. Lee, "Changing Selves, Changing Society." On "change agents," see Kenneth Benne, "Leaders Are Made, Not Born."

112. Lee, "Changing Selves, Changing Society." On the history of sensitivity training, see Elisabeth Lasch-Quinn, *Race Experts: How Racial Etiquette, Sensitivity Training, and New Age Therapy Hijacked the Civil Rights Revolution*.

113. Carl Rogers, "The Process of the Basic Encounter Group."

114. Particularly Carl Rogers, *On Encounter Groups*.

115. The hierarchy of human needs was developed by Abraham Maslow, who along with Carl Rogers is considered a founder of humanistic psychology. Maslow, *Motivation and Personality*; Maslow, "Self-Actualizing People: A Study of Psychological Health."

116. On schools, see Carl Rogers, "A Plan for Self-Directed Change in an Educational Institute." On the Esalen Institute, see Jeffrey Kripal, *Esalen: American and the Religion of No Religion*.

117. Barbara Cruikshank, *The Will to Empower: Democratic Citizens and Other Subjects*; Herman, *The Romance of American Psychology*.

118. On the shift to seeing the self-actualization of "desires" as a crucial element of consumption, and marketing, see Adam Arvidsson, "On the 'Pre-History of the Panoptic Sort': Mobility in Market Research"; and R. Ziff, "The Role of Psychographics in the Development of Advertising Strategy and Copy."

119. Arlie Hochschild, *The Managed Heart: Commercialization of Human Feeling*; Antonio Negri, "Value and Affect"; Kathi Weeks, "Life within and against Work: Affective Labor, Feminist Critique, and Post-Fordist Politics." On sensitivity training, see Lasch-Quinn, *Race Experts: How Racial Etiquette, Sensitivity Training, and New Age Therapy Hijacked the Civil Rights Revolution*.

120. Carol Downer, "What Makes the Feminist Women's Health Center 'Feminist'?"

121. Carol Downer, Lorraine Rothman, and Eleanor Snow, "F.W.H.C. Response."

122. Judy Leste, Shannon Bennett, Cathie Pascoe, Linda Aldous, Joanne Cline, Lorna Rocha, Sue Keeler, Lorey Bonante, Dianne Sultana, Terri Greenberg, and Zoe Tafoya, "Calling It Quits."

123. This was true of an Atlanta clinic. Wendy Simonds, *Abortion at Work: Ideology and Practice in a Feminist Clinic*. The Santa Cruz Women's Health Center made perhaps the most extensive effort to meld the labor in clinics with the protocols of self help, publishing a series of eight articles on "collective process," some of which also explicitly drew on human relations techniques, such as using "strokes," a technique for giving positive "feedback" while working together. Nonetheless, two years after its series on collective process ended, the Santa Cruz clinic lost access to county revenue and had to lay off twelve of its seventeen paid workers, deciding it was best to pay five workers a livable wage. See series on the collective process running from March 1977 to December 1978 in the *Santa Cruz Women's Health Center Newsletter*, as well as the March 1981 issue, which discusses the collapse of the collective organization, and the March 1982 issue, which discusses the organization's antiracist deliberations and workshops.

124. Jan Thomas and Mary Zimmerman, "Feminism and Profit in American Hospitals: The Corporate Construction of Women's Health Centers."

125. Carol Downer and Colleen Wilson were charged and then acquitted. The police were hoping to catch the women performing a menstrual extraction, and when they failed they fell back on the ludicrous yogurt charge. Wilson plea-bargained on her more complicated charge. The events helped to make Downer a heroine of the movement. Downer, interview, October 24, 1999; Chico Feminist Women's Health Center, "Feminist Women's Health Center Chronology." The open, flexible structure of radical feminist groups made them easy to infiltrate. On the infiltration of radical feminists cells by Trotskyites, see Echols, *Daring to Be Bad*, 214.

126. Women's Community Health Center, Cambridge, Annual Report.

127. Edith Butler, "The First National Conference on Black Women's Health Issues,"

in *Women's Health: Readings on Social, Economic, and Political Issues*, edited by Nancy Worcester and Marianne Whatley (Dubuque, Iowa: Kendall, 1988), 37–41; Byllye, "Breathing Life into Ourselves," 4–10; Deborah Grayson, "'Necessity Was the Midwife of Our Politics': Black Women's Health Activism in the 'Post'-Civil Rights Era (1980–1996)," 131–48; Susan Smith, *Sick and Tired of Being Sick and Tired*.

128. Gwen Braxton, "Self Help is Self-Healing," 6, 7, 12.

129. Avery, "Breathing Life into Ourselves."

130. See Grayson, "'Necessity Was the Midwife of Our Politics'"; and Jael Silliman et al., *Undivided Rights: Women of Color Organize for Reproductive Justice*.

131. On these circuits of self help see chapter 4 in this book; Vin-Kim Nguyen, "Antiretroviral Globalism, Biopolitics, and Therapeutic Citizenship"; and Millie Thayer, "Transnational Feminism: Reading Joan Scott in the Brazilian Sertao."

132. Sonia Alvarez, "Advocating Feminism: The Latin American Feminist NGO 'Boom.'"

Chapter 2. Immodest Witnessing

1. Lorraine Daston and Peter Galison, *Objectivity*.

2. Ibid.

3. Daston, "The Moral Economy of Science."

4. By the 1970s, economists called this kind of highly trained judgment a form of "human capital."

5. See, for example, Kelly Moore, *Disrupting Science: Social Movements, American Scientists, and the Politics of the Military, 1945–1975*.

6. Eric Vettel, *Biotech: The Countercultural Origins of an Industry*.

7. Ibid. On entrepreneurial subjects and technoscience, see Kerry Holden and David Demeritt, "Democratising Science?"; and Tiziana Terranova, "Free Labor: Producing Culture for the Digital Economy."

8. On biomedicine, see my discussion in chapter 1, and Clarke et al., "Biomedicalization."

9. Greene, *Prescribing by Numbers*.

10. Alberto Cambrosio et al., "Regulatory Objectivity and the Generation and Management of Evidence in Medicine."

11. On the engineering approach to biology, see Angela Craeger, Elizabeth Lunbeck, and Norton Wise, *Science without Laws: Model Systems, Cases, Exemplary Narratives*; Hannah Landecker, *Culturing Life: When Cells Became Technologies*; and Philip Pauly, *Controlling Life: Jacques Loeb and the Engineering Ideal in Biology*.

12. I want to acknowledge my own commitments to an understanding of technoscience as world making and not just truth telling. However, this commitment can be politicized in multiple and antagonistic ways, and thus it is worth paying attention to the variety of politicizations, from injunctions to turn science into an engine of capital, to abstract philosophical commitments from theoretical physics, to neoconserva-

tive commitments dismissive of scientific truth telling yet still invested in the performance of technology.

13. On the reassembly of reproduction in this way, see Franklin, *Dolly Mixtures*.

14. Natasha Myers, "Molecular Embodiments and the Body-Work of Modeling in Protein Crystallography." On theorizing affective entanglements as ontology, see Barad, *Meeting the University Halfway*.

15. This is my own paraphrase.

16. Federation of Feminist Women's Health Centers (hereafter, FFWHC), *How to Stay out of the Gynecologist's Office*, 1.

17. See also Adele Clarke and Lisa Moore, "Clitoral Conventions and Transgressions: Graphic Representations in Anatomy Texts, C1900–1991."

18. Haraway, *Modest-Witness@Second-Millennium. Femaleman-Meets-Oncomouse*, 11. For the use of figures more generally, see 8–14.

19. Haraway, "Situated Knowledges"; Sharon Traweek, *Beamtimes and Lifetimes*.

20. On images of nature personified and naked, see Ludmilla Jordanova, *Sexual Visions*; and Katherine Park, "Nature in Person: Renaissance Allegories and Emblems."

21. Steven Shapin and Simon Schaffer, *Leviathan and the Air-Pump*.

22. Haraway, "Situated Knowledges"; Traweek, *Beamtimes and Lifetimes*.

23. Simon Schaffer, "Self Evidence."

24. Kathie Sarachild, a graduate of Harvard, former white civil rights worker in Mississippi, and founding member of New York Radical Women, was one of the earliest architects of feminist versions of consciousness raising. She outlined her program for consciousness raising at the First National Women's Liberation Conference in Chicago, November 27, 1968. Sarachild, "Consciousness-Raising," 145. This essay was first presented as a talk at the First National Conference of Stewardesses for Women's Rights, New York, March 12, 1973.

25. For a published account see FFWHC, *A New View of a Woman's Body*. The women who took part in this study were some of the most involved and longest-standing participants in feminist self help: Carol Downer, Suzanne Gage, Sherry Schiffer, Lorraine Rothman, Frances Hornstein, Lynn Heidelberg, Kathleen Hodge, Lynn Walker, Chris Clear, and Nancy Walker. For more on the clitoral study, see Clarke and Moore, "Clitoral Conventions and Transgressions"; and Nancy Tuana, "Coming to Understand: Orgasm and the Epistemology of Ignorance."

26. See Frances Hornstein, *Lesbian Health Care*. See also Suzanne Gage, *Lesbian Health Activism*.

27. Daston, "The Moral Economy of Science."

28. Sara Ahmed, "Affective Economies."

29. Script for slide show from Lorraine Rothman.

30. Downer, interview, October 24, 1999.

31. For some genealogical reflections on the small group format, see chapter 1.

32. West Coast Sisters, "Self Help Clinic, Part II."

33. Downer, interview, October 24, 1999.

34. Here, I am building on the work of Joan Scott, in particular, "The Evidence of Experience."

35. Betty Friedan, *The Feminine Mystique*, 11.

36. Carol Hanish, "The Personal Is Political."

37. Downer, interview, October 24, 1999.

38. The phrase comes from a talk given in March 1969 and reprinted in Hanish, "The Personal is Political."

39. On the genre, see Brian Norman, "The Consciousness-Raising Document, Feminist Anthologies and Black Women in Sisterhood Is Powerful."

40. Downer, interview, October 24, 1999.

41. Originally published in *Notes from the Third Year* (1970). Reprinted in Pamella Allen, "Free Space," in *Radical Feminism*, edited by Anne Koedt, Ellen Levine, and Anita Rapone, 271–79 (New York: Quadrangle Books, 1973), 275.

42. Puzzle metaphor from Rothman, interview, October 23, 1999.

43. Boston Women's Health Book Collective. *Our Bodies, Ourselves*, 2.

44. FFWHC, *How to Stay out of the Gynecologist's Office*.

45. Suzann Gage, interview, October 25, 1999.

46. My attention here to the "not uncommon" offers a more subtle understanding of the epistemological stakes of vaginal self-exam than in my previous account of this practice. Murphy, "Immodest Witnessing," 115–41.

47. On the professional separation of the mouth from the rest of the body, see Sarah Nettleton. *Power, Pain, and Dentistry*.

48. FFWHC, *How to Stay out of the Gynecologist's Office*, 24.

49. Ibid., 25.

50. This analogy was used in the defense of Carol Downer and Colleen Wilson in the court case that charged them with practicing medicine without a license for applying yoghurt to another woman.

51. Here, I draw on Judith Butler's work, as well as my own reworking of the concept. Butler, *Bodies That Matter*; Murphy, *Sick Building Syndrome and the Problem of Uncertainty*.

52. Butler, *Bodies That Matter*, 2.

53. Susanne Gage, interview. October 25, 1999.

54. For more on the politics of the term *well woman*, see chapter 3.

55. For a more elaborate discussion of unraced feminism, see the previous chapter.

56. This is a frequent move in late twentieth-century liberal racial/antiracist formations, and has canonical expression in the UNESCO statements on race.

57. See chapter 1 for a detailed version of this argument.

58. For a brief genealogy of sex/gender see Murphy, "Sex and Gender."

59. This argument is indebted to theorizations of desire in queer studies that, building on the work of Foucault, historicize desire as evoked in political economies. See, for example, Lisa Rofel, *Desiring China: Experiments in Neoliberalism, Sexuality and Public Culture*.

60. Edward Bernays, *Propaganda*, 75–76. For influential marketing understandings of machines and strategies of desire, see Paul Mazur, *American Prosperity*, and Ernest Dichter, *The Strategy of Desire*. See also William Leach, *Land of Desire*.

61. This process is nicely captured in Rofel, *Desiring China*.

62. See Ziff, "The Role of Psychographics in the Development of Advertising Strategy and Copy"; and Adam Arvidsson, "On the 'Pre-History of the Panoptic Sort.'"

63. Arvidsson, "On the 'Pre-History of the Panoptic Sort.'"

64. In the early 1970s, the crucial debates in this vein of critical thought concerned housework. See, for example, Hodee Edwards, "Housework and Exploitation: A Marxist Analysis"; Silvia Federici, *Wages against Housework*; and Selma James and Mariarosa Dalla Costa, *The Power of Women and the Subversion of the Community*.

65. See Hochschild, *The Managed Heart: Commercialization of Human Feeling*.

66. On affective labor in late twentieth-century capitalism, see Hochschild, *The Managed Heart*; Maurizio Lazzarato, "Immaterial Labor"; Negri, "Value and Affect"; Terranova, "Free Labor"; and Weeks, "Life within and against Work."

67. Feminist technoscience studies, as a academic field, arose in the early 1980s. Thematic arguments for positively valuing affective relations in knowledge production recurred in such influential works as Donna Haraway, "Manifesto for Cyborgs: Science, Technology and Socialist Feminism in the 1980s"; and Evelynn Fox Keller, *A Feeling for the Organism: The Life and Work of Barbara McClintock*.

68. See Haraway, "Situated Knowledges: The Science Question in Feminism and the Privilege of Partial Perspective"; Alison Jagger, "Love and Knowledge: Emotion in Feminist Epistemology"; Evelyn Fox Keller, *A Feeling for the Organism*; and Hilary Rose, "Hand, Brain, and Heart: A Feminist Epistemology for the Natural Sciences." For one Marxian version of the importance of affective labor in capitalism, see Michael Hardt and Antonio Negri, *Empire*, and *Multitude: War and Democracy in the Age of Empire*. For feminist critiques, see Susanne Schultz, "Dissolved Boundaries and 'Affective Labor': On the Disappearance of Reproductive Labor and Feminist Critique in Empire"; Weeks, "Life within and against Work."

69. Rothman, interview, October 23, 1999.

70. FFWHC, *How to Stay out of the Gynecologist's Office*, 24–25.

71. Ibid., 24.

72. "Vaginal Infection Slides (Wet Mount)," in Feminist Women's Health Centers, *Well Woman Health Care in Woman Controlled Clinics*.

73. The women who participated in this study were Suzann Gage, Carol Downer, Karen Grant, Lynn Heidelberg, Kathy Hodge, Frances Hornstein, Margo Miller, Sylvia Morales, and Lorraine Rothman.

74. Rothman, interview. October 23, 1999.

75. FWHC, "Self-Help Study: Observing Changes in the Menstrual Cycle."

76. Sylvia Morales, a professional filmmaker, captured the close-up images of the cervix for the Menstrual Cycle Study that were published in Federation of Feminist Women's Health Centers, *A New View of A Woman's Body*. She also worked with the LA

Feminist Women's Health Center in making the vaginal self-exam film *A New View of Myself* (1975) and later became famous for her documentary *Chicana* (1979).

77. The omission of social location was purposeful. "Clinic Record," in Feminist Women's Health Center, *Well Woman Health Care in Woman Controlled Clinics*.

78. "Self-Help Study: Observing Changes in the Menstrual Cycle."

79. On reproduction as mechanized, see Robbie Davis-Floyd, *Birth as an American Rite of Passage*; and Emily Martin, *The Woman in the Body*.

80. Waldby, "Stem Cells, Tissue Cultures and the Production of Biovalue," 310. See also Franklin's development of the term *biowealth*; and Thompson, *Making Parents*.

81. Patricia Hill Collins, *Black Feminist Thought: Knowledge, Consciousness, and the Politics of Empowerment*; Sandra Harding, "'Strong Objectivity' and Socially Situated Knowledge"; Harding, "Subjectivity, Experience, and Knowledge: An Epistemology from/for the Rainbow Coalition Politics"; Nancy Harstock, "The Feminist Standpoint: Developing the Ground for a Specifically Feminist Historical Materialism"; Dorothy Smith, "Some Implications of a Sociology for Women," and "Women's Perspective as a Radical Critique of Sociology." Sandoval, *Methodology of the Oppressed*, and "New Sciences: Cyborg Feminism and the Methodology of the Oppressed."

82. Harding, "Subjectivity, Experience, and Knowledge," 124.

83. Haraway, "Situated Knowledges."

84. Ibid., 191.

85. Haraway, "The Virtual Speculum for a New World Order."

86. Katz, "On the Grounds of Globalization." See the introduction of this book for its use of topology.

87. This phrase is from Hasper, Heckert, and Schnitger, interview, November, 1999.

Chapter 3. Pap Smears, Cervical Cancer, and Scales

1. This chapter owes a huge debt to Adele Clarke for her help and her work on the history of the Pap smear, reproductive sciences, and biomedicalization, which informs every aspect of this argument. Moreover, for many years she wrote the entry for cervical health in *Our Bodies, Ourselves*. Her work has been an example of the kind of double vision I've tried to create here—the critical examination of what one cannot live without, including feminism, yet through a feminist lens.

2. In terms of periodization, and like the rest of the book, while I recognize that biomedicine crystallized in the 1980s, I still use the term for the specific tendrils of biomedical practice that congealed in the 1970s.

3. Maren Klawiter has developed the concept "cultures of action" to characterize the styles of activism within a social movement. Here, I build on her work by emphasizing, not culture, but the different ways of mapping and spatializing political projects. Klawiter, "Racing for the Cure, Walking Women, and Toxic Touring: Mapping Cultures of Action within the Bay Area Terrain of Breast Cancer."

4. For example, Sallie Marston, "The Social Construction of Scale."

5. I hope that by critically emphasizing the different scales that feminists partici-pated and intervened in, I can avoid an analysis that simply assigns degrees of radical-ness and conformity to different feminisms. I am not setting out to pick apart these feminist interventions for their inadequacy, oversights, and partiality—for working in the wrong place with the wrong map—or insisting that a corrected feminism, which we in the present can better see, would have a perfected cartography and would have folded itself perfectly. Instead, I want to hold these feminist interventions at a critical distance in order to historicize how their participation in technoscience was shaped by, constitutive of, and altered in differently politicized scales that were often en-tangled with each other. In so doing, this chapter seeks to offer its own imperfect critical map, of a multiplicity of feminist projects entangled with a changing stratified biomedicine.

6. There are even higher estimates in the published literature that promote screen-ing, but firm statistics for the United States are rare. On reduction in cancer mortality among white women from 1947 to 1984, see S. S. Devesa, D. T. Silverman, J. L. Young, et al., "Cancer Incidence and Mortality Trends among Whites in the United States, 1947–1984." A good source for statistics is L. A. Ries, C. L. Kosary, B. F. Hankey, et al., *Seer Cancer Statistics Review 1973–1995*.

7. Deborah Kuhn McGregor, *From Midwives to Medicine: The Birth of American Gyne-cology*.

8. Laura Briggs, "The Race of Hysteria: 'Overcivilization' and the 'Savage' Woman in Late Nineteenth-Century Obstetrics and Gynecology"; McGregor, *From Midwives to Medicine*; Carroll Smith-Rosenberg, *Disorderly Conduct: Visions of Gender in Victorian American*.

9. Fredrick Hoffman, *Cancer and Diet, with Facts and Observations of Related Sub-jects*; Hoffman, "The Cancer Problems and Research, Delivered before the Canadian Public Health Association"; Helen Kirchhoff and R. H. Rigdon, "Frequency of Cancer in the White and Negro: A Study Based upon Necropsies"; Raymond Pearl and Agnes Bacon, "The Racial and Age Incidence of Cancer and of Other Malignant Tumors"; R. H. Rigdon, Helen Kirchhoff, and Mary Lee Walker, "Frequency of Cancer in the White and Colored Races as Observed at the Autopsy between 1920 and 1949 at the Medical Branch."

10. Hoffman, "The Cancer Problems and Research, Delivered before the Canadian Public Health Association."

11. Ibid. See also Evelynn Hammonds research on ovarian tumors, personal com-munication; Keith Wailoo, *Dying in the City of Blues: Sickle Cell Anemia and the Politics of Race and Health*; and Wailoo, *When Cancer Crossed the Color Line*.

12. On the gendering of early cancer prevention, see Kirsten Gardner, *Early Detec-tion: Women, Cancer, and Awareness Campaigns in the Twentieth-Century United States*. On the history of cancer in the United States, see, for example, Lester Breslow, *History*

of Cancer Control in the U.S., 1946–1971; Barron Lerner, *The Breast Cancer Wars: Hope, Fear, and the Pursuit of a Cure in Twentieth-Century America*; James Patterson, *The Dread Disease: Cancer and Modern American Culture*; and Walter Ross, *Crusade: The Official History of the American Cancer Society*.

13. Wailoo, *When Cancer Crossed the Color Line*.

14. For an extended version of this argument, see Murphy, "Uncertain Exposures and the Privilege of Imperception: Activist Scientists and Race at the U.S. Environmental Protection Agency" in *Landscapes of Exposure: Knowledge and Illness in Modern Environments*.

15. On cellularity and time, see Landecker, *Culturing Life*.

16. Clarke, *Disciplining Reproduction: Modernity, American Life Sciences and the "Problems of Sex."*

17. George Papanicolaou, "The Sexual Cycle in the Human Female as Revealed by Vaginal Smears." On the history of the various classification schemes used in Pap smear screening, see Adele Clarke and Monica Casper, "From Simple Technology to Complex Arena: Classification of Pap Smears, 1917–1990."

18. For a more thorough discussion of my use of materialization, see Murphy, *Sick Building Syndrome and the Problem of Uncertainty*.

19. On the concept of implicated actor, see Adele Clarke and Theresa Montini, "The Many Faces of Ru486: Tales of Situated Knowledges and Technological Contestations"; Clarke, "From Grounded Theory to Situated Analysis: What's New? Why? How?" On the politics of the one-size-fits-all model in contraception research, see Nelly Oudshoorn, "The Decline of the One-Size-Fits-All Paradigm, or, How Reproductive Scientists Try to Cope with Postmodernity."

20. Papanicolaou and Herbert Traut, *Diagnosis of Uterine Cancer by the Vaginal Smear*. On the history of Pap smear classification, see Clarke and Casper, "From Simple Technology to Complex Arena."

21. Monica Casper and Adele Clarke, "Making the Pap Smear into the 'Right Tool' for the Job: Cervical Cancer Screening in the USA, Circa 1940–95."

22. George Papanicolaou and Herbert Traut, "The Diagnostic Value of Vaginal Smears in Carcinoma of the Uterus," 194.

23. Ibid., 205.

24. Crusade Ross, *Crusade: The Official History of the American Cancer Society*.

25. Casper and Clarke, "Making the Pap Smear into the 'Right Tool' for the Job."

26. The annual gynecological exam joined the annual physical as a mid-twentieth-century, risk-calculating mode of medical care as part of the rise of medical insurance. See Audrey Davis, "Life Insurance and the Physical Examination."

27. For an early discussion of the "well-woman" by an American gynecologist, see A. Clair Siddall, "The Presumably Well Woman."

28. Feminist Women's Health Center, *Well Woman Health Care in Woman Controlled Clinics*; Morgan, "Contested Bodies, Contested Knowledges."

29. Clarke and Casper, "From Simple Technology to Complex Arena."

30. Lerner, *The Breast Cancer Wars*.

31. Frederick Hohmeister, "The Gynecologic Examination," 1181.

32. A. Clair Siddall, "The Gynecologist's Role in Comprehensive Medical Care," 660.

33. Ibid., 658.

34. A. Clair Siddall, "Preventative Medicine in Gynecologic Practice."

35. J. L. Marx, "The Annual Pap Smear: An Idea Whose Time Has Gone?"

36. Wendy Mitchinson identifies 1940 as the approximate point when half of all births were in hospitals. Mitchinson, *Giving Birth in Canada, 1900–1950*; Elizabeth Watkins, *On the Pill: A Social History of Oral Contraceptives, 1950–1970*.

37. On the racialized distribution of privilege with special attention to California, see George Lipsitz, *The Possessive Investment in Whiteness: How White People Benefit from Identity Politics*. See also the detailed discussion in chapter 1.

38. Feminist Women's Health Center, *Well Woman Health Care in Women Controlled Clinics*.

39. For a detailed discussion of protocol feminism, see chapter 1.

40. Vaginal self-exam is discussed in detail in chapter 2.

41. "Taking a Pap Smear," in Feminist Women's Health Center, *Well Woman Health Care in Women Controlled Clinics*.

42. See chapter 1 for a detailed argument on unraced feminism and the problem-space of the clinic.

43. On the history of informed consent and cancer in the postwar period, but without an attention to feminism, see Barron Lerner, "Beyond Informed Consent: Did Cancer Patients Challenge Their Physicians in the Post–World War II Era?"

44. Allan Brandt, "Behavior, Disease, and Health in the Twentieth-Century United States: The Moral Valence of Individual Risk."

45. See chapter 1. Lawrence Cohen develops this term in Lawrence Cohen, "Operability, Bioavailability, and Exception."

46. Clarke et al., "Biomedicalization."

47. This materialization of the vaginal ecology is discussed in detail in the previous chapter. See Federation of Feminist Women's Health Centers, *A New View of a Woman's Body*.

48. On individually variegated normality, see chapter 2.

49. Federation of Feminist Women's Health Centers, *A New View of a Woman's Body*.

50. Mimeographed script from Los Angeles feminist self help traveling slide show, provided by Lorraine Rothman, ca. 1974.

51. Ibid.

52. In Canada, the Pap smear was offered in a provincially run, federally supported heath care system. In Barbados, Pap smears were not available widely, and tended to be distributed through the Barbados Family Planning Association. Bruce Barron and Ralph Richart, "An Epidemiologic Study of Cervical Neoplastic Disease: Based on a

Self-Selected Sample of 7,000 Women in Barbados, West Indies"; H. W. Vaillant, G. T. Cummins, and R. M. Richart, "An Island-Wide Screening Program for Cervical Neoplasia in Barbados."

53. For example, see Ann Carson et al., *Choosing a Pap Smear Lab: A Guide for the Health Care Provider*; Mary McNamara, "Pap Smears: Testing the Tests."

54. The figure of women in the biomedical assembly line is very much like the figure of women in the integrated circuit analyzed in the 1980s by Rachel Grossman, "Woman's Place in the Integrated Circuit"; and Donna Haraway, "A Cyborg Manifesto: Science, Technology, and Socialist-Feminism in the Late Twentieth Century."

55. See, for example, Walt Bogdanich, "Lax Laboratories: The Pap Test Misses Much Cervical Cancer through Labs' Errors."

56. On the process of desegregation and its conjuncture with Medicaid, see Michael Byrd and Linda Clayton, *An American Health Dilemma: Race, Medicine, and Health Care in the United States, 1900–2000*.

57. While Canada does not, yet, have privatized medicine, the cost-benefit analysis also made tremendous inroads there, as well as in Britain, though with a greater tendency to include "benefits to society" in calculations than in the United States. For an excellent account of the epistemological content of this shift, see Clarke et al., "Biomedicalization."

58. The range of costs/benefit analyses done in the United States, and their comparison to Britain, which tended to factor "social benefit" into their calculations, was reviewed in an article that helped to spark a debate over the efficiency that occurred in the mass media as well as medical communities. Anne Marie Foltz and Jennifer Kelsey, "The Annual Pap Test: A Dubious Policy Success."

59. New England Journal of Medicine, "Editorial: Papanicolaou Testing: Are We Screening the Wrong Women?"

60. K. Fidler, D. Boyes, and A. J. Worth, "Cervical Cancer Detection in British Columbia: A Progress Report," 402–3.

61. Leopold Koss, "The Attack on the Annual 'Pap Smear,'" 182.

62. J. R. Walton et al., "Cervical Cancer Screening Programs."

63. O. Schmidt, "Cervical Cancer Screening Program: The Sogc's View."

64. R. S. Cooper and R. David, "The Biological Concept of Race and Its Application to Public Health and Epidemiology"; Steven Epstein, *Inclusion: The Politics of Difference in Medical Research*; Nancy Krieger, "Racism, Sexism, and Social Class: Implications for Studies of Health, Disease, and Well-Being"; Nancy Krieger, "Refiguring 'Race': Epidemiological, Racialized Biology, and Biological Expressions of Race Relations"; Report of the President's Cancer Panel, "The Meaning of Race in Science: Considerations for Cancer Research"; D. Williams, "Race/Ethnicity and Socioeconomic Status: Measurement and Methodological Issues"; D. Williams, R. Lavizzo-Mourey, and R. C. Warren, "The Concept of Race and Health Status in American."

65. Goldberg, *The Racial State*. See also Foucault's formulation in Foucault, *Society Must Be Defended*.

66. For examples that attempt to theorize racial governmentality beyond Foucault's own formulation, see Roderick Ferguson, "Of Our Normative Strivings: African American Studies and the Histories of Sexuality"; David Theo Goldberg, *Racial Subjects: Writing on Race in America*; and Lisa Lowe, "The Worldliness of Intimacy."

67. Troy Duster, "Feedback Loops in the Politics of Knowledge Production."

68. Etienne Balibar, "Is There a 'Neo-Racism'?"; Paul Gilroy, "One Nation under a Groove: The Cultural Politics of 'Race' and Racism in Britain"; Pierre-Andre Taguieff, "The New Cultural Racism in France."

69. For an excellent discussion of the work of race in epidemiology, see Janet Shim, "Constructing 'Race' across the Science-Lay Divide: Racial Formation in the Epidemiology and Experience of Cardiovascular Disease."

70. V. Carpenter and B. Colwell, "Cancer Knowledge, Self-Efficacy, and Cancer Screening Behaviors among Mexican-American Women"; C. E. Ross, J. Mirowsky, and W. C. Cockerham, "Social Class, Mexican Culture, and Fatalism: Their Effects on Psychological Distress."

71. This correlation between blacks and spirituality has intensified in recent years. R. W. Denniston, "Cancer Knowledge, Attitudes, and Practices among Black Americans"; D. R. Lannin et al., "Influence of Socioeconomic and Cultural Factors on Racial Differences in Late-Stage Presentation of Breast Cancer."

72. Vancouver Women's Health Collective, *A Feminist Approach to Pap Tests*, 20.

73. See, for example, the work of Fredrick Hoffman on cancer: Hoffman, *Cancer and Diet, with Facts and Observations of Related Subjects*, and "The Cancer Problems and Research, Delivered before the Canadian Public Health Association."

74. Here I am again rephrasing a question posed by Joan Scott's work. She asks why historians treat experience as evidence that explains rather than understanding experience as historically produced and thus in need of explanation. The same might be said of gender as a category of analysis. Scott, "The Evidence of Experience."

75. Vancouver Women's Health Collective, *A Feminist Approach to Pap Tests*.

76. These were the three constituencies emphasized in the booklet. British feminist critics tended to emphasize class, and the work of the Boston Women's Health Collective emphasized discrepancies between black and white mortality.

77. Vancouver Women's Health Collective, *A Feminist Approach to Pap Tests*, 15.

78. Ibid.

79. Ibid.

80. Peggy Pascoe, *Relations of Rescue: The Search for Female Moral Authority in the American West, 1874–1939*.

81. John Sung, Ernest Alema-Mensah, and Daniel Blumenthal, "Inner-City African American Women Who Failed to Receive Cancer Screening Following a Culturally-Appropriate Intervention: The Role of Health Insurance"; John Sung et al., "Cancer Screening Intervention among Black Women in Inner-City Atlanta: Design of a Study." For anthropological work on the culture of Pap smear use by black women in Atlanta, see Jessica Gregg and Robert Curry, "Explanatory Models for Cancer among

African-American Women at Two Atlanta Health Centers: The Implications for a Cancer Screening Program."

82. Door-to-door dissemination of reproductive services has a longer history in family planning within both colonial and postcolonial projects funded by the United States. For more on door-to-door dissemination, see chapter 4. See also, Laura Briggs, *Reproducing Empire: Race, Sex, Science and U.S. Imperialism in Puerto Rico.*

83. Abortions, according to a Canadian law overturned in 1988, could only be performed legally and paid for by the state in an accredited hospital.

84. AMA, "Physicians by Gender (Excluding Students)."

85. Calculations based on data provided in Leon Bouvier, "Doctors and Nurses: A Demographic Profile."

86. Nancy Worcester and Mariamne Whatley, "The Response of the Health Care System to the Women's Health Movement: The Selling of Women's Health Centers," 19.

87. Nancy Worcester and Mariamne Whatley, "The Role of Technology in the Co-optation of the Women's Health Movement: The Case of Breast Cancer Screening," 187.

88. Worcester and Whatley, "The Response of the Health Care System to the Women's Health Movement," 21.

89. Ibid., 23.

90. Amy Fairchild, *Science at the Borders: Immigrant Medical Inspection and the Shaping of the Modern Industrial Labor Force*; Eithne Luibhéid, *Entry Denied: Controlling Sexuality at the Border*; Nayan Shah, *Contagious Divides: Epidemics and Race in San Francisco's Chinatown*; Alexandra Minna Stern, "Buildings, Boundaries, and Blood: Medicalization and Nation-Building on the U.S.-Mexico Border, 1910–1930."

91. Sarah Lochlann Jain, "Sentience and Slaver: The Struggle over the Short-Handled Hoe"; Natalia Molina, *Fit to Be Citizens? Public Health and Race in Los Angeles, 1879–1939*; Linda Nash, *Inescapable Ecologies: A History of Environment, Disease, and Knowledge*; Shah, *Contagious Divides*; Alexandra Minna Stern, *Eugenic Nation: Faults and Frontiers of Better Breeding in Modern America.*

92. On how availability relates to biomedicine and citizenship, see Cohen, "Operability, Bioavailability, and Exception." See also chapter 1.

93. Paul Farmer, *Aids and Accusation: Haiti and the Geography of Blame.*

94. Byrd and Clayton, *An American Health Dilemma.*

95. By necropolitics, I mean the governing of death, its distributions, forms, and likelihoods. For a detailed discussion of necropolitics, see chapter 1.

96. See Evelynn Hammonds, "Missing Persons: African American Women, Aids and the History of Disease."

97. Barbara Hulka, "Motivation Technics in a Cancer Detection Program: Utilization of Community Resources." See also Pierre Audet-Lapointe, "Detection of Cervical Cancer in a Women's Prison."

98. Audet-Lapointe, "Detection of Cervical Cancer in a Women's Prison"; R. E. Martin, "A Review of a Prison Cervical Cancer Screening Program in British Columbia"; A. J. Pereyra, "The Relationship of Sexual Activity to Cervical Cancer: Cancer of the Cervix in a Prison Population."

99. Andrea Lewis and Beverly Smith, "Looking at the Total Picture: A Conversation with Health Activist Beverly Smith," 177.

100. Ibid., 179.

101. Audre Lorde, *A Burst of Light*, 116. See also Audre Lorde, *The Cancer Journals*.

102. Davis, "Sick and Tired of Being Sick and Tired," 23.

103. Ibid., 25.

104. For an extended discussion on the Combahee River Collective and simultaneous oppressions, see chapter 1. Combahee River Collective, "Combahee River Collective Statement"; also reprinted as "A Black Feminist Statement" in Gloria Hull, Patricia Bell-Scott, and Smith Barbara, *All the Women Are White, All the Blacks Are Men, but Some of Us Are Brave: Black Women's Studies*.

105. Simon Herrington, "Do HPV-Negative Cervical Carcinomas Exist?—Revisited"; Jan Walboomers et al., "Human Papillomavirus Is a Necessary Cause of Invasive Cervical Cancer Worldwide."

106. See, for example, M. Hutchinson and J. Morales, "Population-Based Study of Human Papillomavirus Infection and Cervical Neopolasia in Rural Costa Rica"; B. Nonnenmacher, N. Hubbert, and R. Kirnbauer, "Serologic Response to Human Papillomavirus Type 16 Virus-Like-Particles in HPV-16 DNA-Positive Invasive Cervical Cancer and Cervical Intraepithelial Neoplasia Grade III Patients and Controls from Colombia and Spain"; W. C. Reeves, L. A. Brinton, and M. Garcia, "Human Papillomavirus Infection and Cervical Cancer in Latin America"; Grazyna A. Stanczuk et al., "Typing of Human Papillomavirus in Zimbabwean Patients with Invasive Cancer of the Uterine Cervix"; Torroella-Kouri et al., "HPV Prevalence among Mexican Women with Neoplastic and Normal Cervixes"; L. L. Villa and E. L. Franco, "Epidemiologic Correlates of Cervical Neoplasia and Risk of Human Papillomavirus Infection in Asymptomatic Women in Brazil"; and Williamson et al., "Typing of Human Papillomaviruses in Cervical Carcinoma Biopsies from Cape Town."

107. R. S. Baile et al., "Trends in Cervical Cancer Mortality in South Africa"; Amie Bishop et al., "Cervical Dysplasia Treatment: Key Issues for Developing Countries"; Amie Bishop et al., "Cervical Cancer: Evolving Prevention Strategies for Developing Countries"; J. C. Gage, C. Ferreccio, and M. Gonzales, "Follow-up Care of Women with Abnormal Cytology in a Low-Resource Setting"; Jessica Gregg, *Virtually Virgins: Sexual Strategies and Cervical Cancer in Recife Brazil*; Eduardo César Lazcano-Ponce et al., "Cervical Cancer Screening in Developing Countries: Why Is It Ineffective? The Case of Mexico." On the importance of medical supplies in reproductive health, see Ana Teresa Ortiz, "'Bare-Handed' Medicine and Its Elusive Patients: The Unstable Construction of Pregnant Women and Fetuses in Dominican Obstetrics Discourse."

108. Women and Development Unit, "Demystifying and Fighting Cervical Cancer."

109. World Health Organization, *World Cancer Report*.

110. Judith Wasserheit, "The Significance and Scope of Reproductive Tract Infections among Third World Women."

111. Ibid.

112. Sabine Lang, "The NGO-ization of Feminism." See also Jael Silliman, "Expanding Civil Society: Shrinking Political Spaces. The Case of Women's NGOs."

113. Alvarez, "Latin American Feminisms Go Global."

114. For a critical insider's analysis, see Petchesky, *Global Prescriptions*.

115. Other examples are the Women's Environment and Development Organization, and the Center for Women's Global Leadership.

116. For more details on this point, see chapter 4.

117. It is notable that this statement was published as late as 2004. IWHC, *Twenty Years, One Goal: IWHC's 20th Anniversary Report*, 2.

118. Cruikshank, *The Will to Empower*.

119. The conference papers were published as Adrienne Germain et al., *Reproductive Tract Infections: Global Impact and Priorities for Women's Reproductive Health*.

120. Adrienne Germain and Jane Ordway, *Population Control and Women's Health: Balancing the Scales*, 1.

121. Ibid.

122. IWHC and WAND, *Challenging the Culture of Silence: Building Alliances to End Reproductive Tract Infections*.

123. Ibid., 9.

124. Ibid., 13.

125. Ibid., 19.

126. Ibid., 32. See also Susan Bell, "Sexual Synthetics: Women, Science, and Microbicides."

127. IWHC and WAND, *Challenging the Culture of Silence*, 37.

128. While a vision of the campaign was developed, it was not actually undertaken.

129. For accounts of Andaiye's political work, see David Scott, "Counting Women's Caring Work: An Interview with Andaiye"; and Alissa Trotz, "Red Thread: The Politics of Hope in Guyana."

130. In the 1970s, Selma James developed cutting critiques of the work of sexed unpaid labor in formations of capital, extending her analysis "left of" the anticolonial black Marxism of her partner, the Trinidadian Marxist scholar C. L. R. James. Carol Boyce Davis develops the trope of "left of Marx" to describe the work of Marxist feminist black Left in *Left of Karl Marx: The Political Life of Black Communist Claudia Jones*.

131. Women and Development Unit, "Demystifying and Fighting Cervical Cancer," 31.

132. Ibid., 33.

133. Ibid., 34.

134. Ibid.

135. Ibid., 35.

136. The campaign did not get larger support and never got off the ground. However, a later pan-Caribbean campaign was undertaken in the early 1990s. D. B. Barnett, "Cervical Cancer Screening Programs: Technical Cooperation in the Caribbean."

137. For a Science and Technology Studies account of this conference, see Halfon, *The Cairo Consensus*.

138. "Women's Voices '94: A Declaration on Population Policies."

139. For a critical account of these developments, see Dennis Hodgson and Susan Watkins, "Feminists and Neo-Malthusians: Past and Present Alliances."

140. Petchesky, *Global Prescriptions: Gender, Health and Human Rights*.

141. See the discussion of Lawrence Summers in the introduction, as well as see Michelle Murphy, "Distributed Reproduction."

142. On historicizing "power," see chapter 1.

143. Transcript of speech, IWHC fundraising gala 2006 in New York City, Kati Marton, Remarks by IWHC board chair, January 19, 2006.

144. Accompanying his wife, Jennifer Oppenheimer, the IWHC gala marked the first time in fifty years since a major diamond trading company executive visited the United States. Rosalind Kainyah is executive director of corporate communications for the Diamond Trading Company, the sales and marketing arm of the De Beers Group. Also speaking at the gala was the executive sales director, Varda Shine. Transcript of speech, IWHC fundraising gala 2007 in New York City. Remarks by Rosalind Kainyah, De Beers Group (2007).

145. Hartmann, *Reproductive Rights and Reproductive Wrongs*.

146. Amy Dalton, "Rediscovering the Arusha Declaration and Abandoning the NGO Model."

147. Sarah White, "Thinking Race, Thinking Development."

148. See, for example, the analysis in Petchesky, *Global Prescriptions: Gender, Health and Human Rights*.

149. R. Sankaranarayanan et al., "Visual Inspection of the Uterine Cervix after the Application of Acetic Acid in the Detection of Cervical Carcinoma and Its Precursors"; D. C. Slawson, J. H. Bennett, and J. M. Herman, "Are Papanicolaou Smears Enough? Acetic Acid Washes of the Cervix as Adjunctive Therapy: A Harnet Study"; University of Zimbabwe / JHPIEGO Cervical Cancer Project, "Visual Inspection with Acetic Acid for Cervical-Cancer Screening: Test Qualities in a Primary-Care Setting."

150. World Health Organization, *World Cancer Report* (2005).

151. Forbes Group, "Plotting the Future of Cytotechnology: An Environmental Analysis of the Driving Forces of Cytology."

152. Keith Wailoo et al., *Three Shots at Prevention: The HPV Vaccine and the Politics of Medicine's Simple Solutions*.

153. The political storm this caused led to both the state legislature undoing the law and to Merck backing away from this tactic in the United States.

154. This is the result of a 1996 immigration law that requires any vaccine recommended, even if not required, by the CDC.

155. Prasun Chaudhuri, "Killer at Large," *Telegraph*, November 21, 2005.

156. "Guard Yourself Commercial," www.youtube.com. 2007.

Chapter 4. Traveling Technology

1. Richard Soderstrom, "Menstrual Regulation Technology," 60.

2. On multiplicity and historical ontology, see Annemarie Mol, *The Body Multiple: Ontology in Medical Practice*; Mol and Law, *Complexities*; and Murphy, *Sick Building Syndrome and the Problem of Uncertainty*.

3. See discussion in the introduction to this book, as well as Murphy, "The Economization of Life."

4. Margaret Sanger called the problem of birth and its control the "pivot of civilization." In her formulation, however, this pivot is rendered as an a priori human condition, rather than a historicizable biopolitical configuration.

5. Greenhalgh and Winckler, *Governing China's Population: From Leninist to Neoliberal Biopolitics*; Tyrene White, *China's Longest Campaign: Birth Planning in the People's Republic, 1949–2005*.

6. Pang-chung Ch'eng, Ch'in-jui Yang, and Kuo-ying Fang, "Use of the 'Fire Cup' Type Uterine Suction Bottle for Induced Abortion," *Chinese Journal of Nursing* 2, no. March (1966); Leo Orleans, "Abortion"; Jun-min P'an, "Application of the Simple Negative Pressure Bottle and Its Removable Hand-Operated Suction Equipment"; Hung-chao Sung, "A Revolutionized Mind Is the Key to Success in Research Work"; Y. T. Wu and H. C. Wu, "Suction in Artificial Abortion: Report of 300 Cases"; Yuan-t'ai Wu and Ling-mei Shang, "Clinical Application of the Negative Pressure Bottle."

7. Wu and Wu, "Suction in Artificial Abortion: Report of 300 Cases." See republication of an image from this in Redstockings, *Feminist Revolution*.

8. Stephanie Caruana, "The Complete Facts on Abortion."

9. Jan BenDor, "Karman as 'Hero': Abortion Innovator Exhibits Sexism"; Philip Goldsmith, "Interview with Harvey Karman"; Helen Koblin, "Vaginal Politics, Dr. Karman, Abortionist."

10. The most common type of medical abortion technique in California until the 1970s was still dilation and curettage (D&C), which, though not complicated, was a form of gynecological surgery requiring anesthesia. Curettage, a scraping procedure using a spoonlike instrument, and dilation, a method of inserting successively larger rods into the opening of the uterus, were both late nineteenth-century surgical techniques originating in France. Before D&C came to America, abortions were primarily achieved through herbal medicines or douching solutions that were sold in America in a large black market or passed down as popular knowledge. Andrea Tone, *Devices and Desires: A History of Contraception in America*. The professionalization of American medicine in the mid-nineteenth century, and gynecology soon after, necessitated

distinguishing between legitimate forms of fertility control, which physicians performed, and illegitimate forms, distributed and practiced by nonphysicians. Dilation and curettage was the dominant method provided by physicians, openly practiced until the mid-twentieth century. This distinction between legitimate and illegitimate abortion methods was, by 1900, inscribed into law in every state. Abortion became a procedure practiced legally only by doctors, at their discretion. Kristin Luker, *Abortion and the Politics of Motherhood*, 15. Yet, abortions, as the historians Kristin Luker and Leslie Reagan have shown, remained quite common in the first half of the twentieth century, estimated to have occurred at a conservative ratio of one abortion (both surgical and nonsurgical) for every five live births. Luker, *Abortion and the Politics of Motherhood*, 49–51; Leslie Reagan, *When Abortion Was a Crime: Women, Medicine, and Law in the United States, 1867–1973*. Total abortions performed exceeded those performed by physicians, and a thriving illegal abortion trade made up the difference. A safely administered D&C required the same sanitary conditions and precautions as other surgical procedures, conditions not easily created in the illegal trade, and thus women risked infection and even death undergoing them. When Downer met Karman, aspiration abortion methods, using vacuum pumps and rigid cannulas, were superseding D&Cs as the favorite method practiced by physicians for pregnancies up to the tenth week. Recent aspirations methods had originated in China and had spread to Japan, Russia, and Eastern Europe in the 1950s. The Cold War worked to slow the spread of this technique into Western Europe, yet by the late 1960s aspiration methods accounted for one-third of abortions in Britain, and by the early 1970s, one study estimated that 81 percent of abortions before twelve weeks' gestation were performed using this method.

11. Harvey Karman, "The Paramedic Abortionist." Harvey Karman and Richard Scotti, "Menstrual Regulation and Early Pregnancy Termination Performed by Paraprofessionals under Medical Supervision," 367.

12. On Jane, see Laura Kaplan, *The Story of Jane: The Legendary Underground Feminist Abortion Service*.

13. The California Therapeutic Abortion Act of 1967 allowed for abortions in an accredited hospital, at up to twenty weeks' gestation, and subject to approval by a panel of doctors. While this law was more open than that of most states, it nonetheless strictly regulated who could perform abortions. Only doctors in hospitals could legally perform abortions, leaving outside the law a wide range of practitioners who had provided abortions before legalization, including nonphysician abortionists such as Karman and feminist health activists such as Downer.

14. Pat Maginnis and Lena Phelan, *The Abortion Handbook for Responsible Women*. See also Ninia Baehr, *Abortion without Apologies: A Radical History for the 1990s*; and Reagan, *When Abortion Was a Crime: Women, Medicine, and Law in the United States, 1867–1973*.

15. Karmen visited Paris in the early 1970s and introduced his abortion method, called the "Karmen Method." The specter of women's ability to perform abortion irre-

spective of the law was directly cited in the 1974 legislation that legalized abortion in France. Melanie Latham, *Regulating Reproduction: A Century of Conflict in Britain and France.*

16. Lorraine Rothman, "Menstrual Extraction," 9 (Los Angeles, n.d., ca. late 1970s), Boston Women's Health Book Collective Archive, file "Menstrual Extraction."

17. Rothman would not reveal what this name meant in our interview. United States Patent 3,828,781, August 13, 1974.

18. Debra Law in *The Proceedings of the Menstrual Extraction Conference*, 34.

19. Shelly Farber, quoted ibid., 13.

20. On the practices of production in open software, see Chris Kelty's work on UNIX in Kelty, *Two Bits: The Cultural Significance of Free Software.*

21. "Woman Controlled Research" (n.d., ca. 1970s), Boston Women's Health Collective Archive, file "Menstrual Extraction."

22. Rothman in *The Proceedings of the Menstrual Extraction Conference*, 12.

23. Women's Choice Clinic, Feminist Women's Health Center, "Menstrual Extraction: The Means to Responsibly Control our Periods," photocopy (Oakland, Calif., n.d., ca. 1970s), Boston Women's Collective Archive, file "Menstrual Extraction."

24. Reimert Ravenholt and D. G. Gillespie, "Maximizing Availability of Contraception through Household Utilization," 7.

25. The ability of USAID to determine actual national population and family-planning programs and services on the ground were attenuated. The organization supplied the funds and materials, but programs and services were delivered by states or by nongovernmental agencies. Bangladesh's National Population Program had little resemblance to USAID's early strategy for the rapid and dramatic dissemination of contraception. By 1976, Bangladesh had adopted a "compensation" program that gave financial incentives to the client, the "motivators" (who could be anyone from a family-planning worker to a local bureaucrat), and clinic staff for sterilizations and IUDs. Incentives included a sari for women and a lungi for men, and what amounted to a week's worth of wages for an unskilled rural laborer. For a critique of these programs, see Hartmann and Standing, *Food, Saris and Sterilization.* For work within the population control industry in support of incentives in Bangladesh, see John Cleland and W. Parker Mauldin, "The Promotion of Family Planning by Financial Payments."

Most of the funding for this program came from USAID, constituting its largest bilateral population commitment to any country. The incentive system became ripe with abuse and was denounced as a form of coercion by many feminists. See the important critiques of Unnayan Bikalper Nitinirdharoni Gobeshona (UBINIG) in Akhter, *Depopulating Bangladesh.* By the end of Ravenholt's tenure, the USAID office was promoting the permanent fix of sterilization as much as it did the pill. In practice, USAID was reliant on both state and nongovernmental family-planning agencies for actual delivery of services in the field.

26. Ravenholt, "Taking Contraceptives to the World's Poor: Creation of USAID's Population/Family Planning Program, 1965–80," 31.

27. Ibid.

28. Ravenholt, "AID's Family Planning Strategy."124.

29. Ravenholt, "The Power of Availability," 2–3.

30. Jane Brody, "Physicians throughout the World Are Studying New, Simple Technique for Terminating Pregnancy." *New York Times*, December 20, 1973, 26.

31. "Killing of Babies Feared in Bengal"; Audrey Menen, "The Rapes of Bangladesh; the Rapes"; Robert Trumbull, "Dacca Raising the Status of Women While Aiding Rape Victims."

32. Nayanika Mookherjee, "Gendered Embodiments: Mapping the Body-Politic of the Raped Woman and the Nation in Bangladesh" and *The Spectral Wound: Sexual Violence, Public Memories and the Bangladesh War of 1971*; Bina D'Costa, *Nationbuilding, Gender and War Crimes in South Asia*; Yasmin Saikia, *Women, War and the Making of Bangladesh*.

33. Exact figures are controversial. Some women used nonmedical abortion methods, and some eighteen clinics performed medical abortions. Trumbull, "Dacca Raising the Status of Women While Aiding Rape Victims." For debate over numbers, see the controversies around Sarmila Bose, *Dead Reckoning: Memories of the 1971 Bangladesh War*.

34. Leonard Laufe, "Menstrual Regulation: International Perspectives."

35. Nayanika Mookherjee, "Available Motherhood: Legal Technologies, 'States of Exception' and the Dekinning of 'War-Babies' in Bangladesh."

36. Ruth Dixon-Mueller, "Innovations in Reproductive Health Care: Menstrual Regulation Policies and Programs in Bangladesh," 129.

37. On the transnational formation of CROs, see Petryna, Lakoff, and Kleinman, *Global Pharmaceuticals: Ethics, Markets, Practices*. On its relation to biocapital, see Rajan, *Biocapital*.

38. PharmalinkFHI is now called Novella Clinical.

39. Thank you to Kaushik Sunder Rajan, Kristen Peterson, Vinh-Kim Nguyen, and Joe Dumit for conversations about this relationship.

40. Laura Brown's attendance at the Hawaii Menstrual Regulation Conference was described by Lolly Hirsch, Untitled, *Monthly Extract*, 8.

41. Rothman in, *The Proceedings of the Menstrual Extraction Conference*, 11.

42. Hirsch, "Practicing Health without a License," *Monthly Extract* (July/August 1977): 3.

43. Quote from Shelly Farber at First National Women Controlled Women's Health Clinic Conference, held in Ames, Iowa, November 1974. Quoted in Hirsch, "Second Women-Controlled Women's Health Center Conference," 2.

44. Ibid.

45. I take the term "responsibilized" from Nikolas Rose, "Governing 'Advanced' Liberal Democracies."

46. Bruno Latour, *Science in Action*.

47. Women's Choice Clinic, Feminist Women's Health Center, "Menstrual Extrac-

tion: The Means to Responsibly Control our Periods," photocopy (Oakland, Calif., n.d., ca. 1970s), Boston Women's Collective Archive, file "Menstrual Extraction."

48. Rothman, "Menstrual Extraction," 7.

49. Abul Barkat and Tawheed Reza Noor, *Comprehensive Reproductive Health Services in Urban Slums: Bangladesh Women's Health Coalition Agargaon Project, Upscaling Innovations in Reproductive Health*; Bonnie Kay, Adrienne Germain, and Maggie Bangser, *The Bangladesh Women's Health Coalition, Quality/Calidad/Qualité*, 2, 4.

Conclusion

1. Sister Song Women of Color Reproductive Health Collective, *Reproductive Justice Briefing Book*, 4.

2. See Murphy, "Distributed Reproduction."

Adorno, Theodore, Else Frenkel-Brunswik, D. F. Levinson, and R. N. Sanford. *The Authoritarian Personality*. New York: Harper, 1950.

African American Women for Reproductive Freedom. "We Remember." In *Still Lifting, Still Climbing: African American Women's Contemporary Activism*. Edited by Kimberly Springer, 38–41. New York: New York University Press, 1999.

Agamben, Giorgio. *Homo Sacer: Sovereign Power and Bare Life*. Translated by Daniel Heller-Roazen. Stanford: Stanford University Press, 1998.

Ahmed, Sara. "Affective Economies." *Social Text* 22, no. 2 (2004): 118–39.

Akhter, Farida. *Depopulating Bangladesh: Essays on the Politics of Fertility*. Dhaka: Narigrantha Prabartana, 1992.

———. "Reproductive Rights: A Critique from the Realities of Bangladeshi Women." *Re/productions*, no. 1 (1998).

Ali, Kamran Asdar. *Planning the Family in Egypt: New Bodies, New Selves*. Austin: University of Texas Press, 2002.

Allen, Pamela. "Free Space." In *Radical Feminism*. Edited by Anne Koedt, Ellen Levine, and Anita Rapone, 271–79. New York: Quadrangle Books, 1973.

Alvarez, Sonia. "Advocating Feminism: The Latin American Feminist NGO 'Boom.'" *International Feminist Journal of Politics* 1, no. 2 (1999): 181–209.

———. "Latin American Feminisms Go Global: Trends of the 1990s, Challenges for the New Millennium." In *Cultures of Politics / Politics of Cultures: Re-Visioning Latin American Social Movements*. Edited by Sonia E. Alvarez, Evelina Dagnino, and Arturo Escobar. Boulder: Westview, 1998.

AMA. "Physicians by Gender (Excluding Students)." *Physician Characteristics and Distribution in the U.S.* 2005.

Amin, Ruhul, Shafiqur Choudhuri, Alemayehu Mariam, and James McCarthy. "Family Planning in Bangladesh, 1969–1983." *International Family Planning Perspectives* 13, no. 1 (1987): 16–20.

Anderson, Warwick. *The Cultivation of Whiteness: Science, Health and Racial Destiny in Australia*. Durham: Duke University Press, 2006.

Anzaldúa, Gloria. *Borderlands / La Frontera: The New Mestiza*. San Francisco: Aunt Lute Books, 1987.

Arvidsson, Adam. "On the 'Pre-History of the Panoptic Sort': Mobility in Market Research." *Surveillance and Society* 1, nos. 3–4 (2004): 456–74.

Audet-Lapointe, Pierre. "Detection of Cervical Cancer in a Women's Prison." *Canadian Medical Association Journal* 104, no. 6 (1971): 509–11.

Avery, Byllye. "Breathing Life into Ourselves: The Evolution of the National Black Women's Health Project." In *The Black Women's Health Book: Speaking for Ourselves*. Edited by Evelyn White, 4–10. Seattle: Seal Press, 1990a.

———. "A Question of Survival / A Conspiracy of Silence: Abortion and Women's Health." In *From Abortion to Reproductive Freedom: Transforming a Movement*. Edited by Marlene Geber Fried, 75–81. Boston: South End Press, 1990b.

———. "Who Does the Work of Public Health." *American Journal of Public Health* 92, no. 4 (2002): 570–74.

Baehr, Ninia. *Abortion without Apologies: A Radical History for the 1990s*. Boston: South End Press, 1990.

Baile, R. S., C. E. Selve, D. Bourne, and D. Bradshaw. "Trends in Cervical Cancer Mortality in South Africa." *International Journal of Epidemiology* 25, no. 3 (1996): 488–93.

Balibar, Etienne. "Is There a 'Neo-Racism'?" In *Race, Nation, Class: Ambiguous Identities*. Edited by Etienne Balibar and Immanuel Wallerstein, 17–28. London: Verso, 1991.

Barkat, Abul, and Tawheed Reza Noor. Comprehensive Reproductive Health Services in Urban Slums: Bangladesh Women's Health Coalition Agargaon Project, Upscaling Innovations in Reproductive Health. Ampang, Malaysia: International Council on Management of Population Programmes, 2000.

Barnett, D. B. "Cervical Cancer Screening Programs: Technical Cooperation in the Caribbean." *Bulletin of the Pan American Health Organization* 30, no. 4 (1996): 409–12.

Barron, Bruce, and Ralph Richart. "An Epidemiologic Study of Cervical Neoplastic Disease: Based on a Self-Selected Sample of 7,000 Women in Barbados, West Indies." *Cancer* 27, no. 4 (1970): 978–86.

Belcher, Oliver, Lauren Martin, Anna Secor, Stephanie Simon, and Tommy Wilson. "Everywhere and Nowhere: The Exception and the Topological Challenge to Geography." *Antipode* 40, no. 4 (2008): 499–503.

Bell, Susan. "Sexual Synthetics: Women, Science, and Microbicides." In *Synthetic Planet: Chemical Politics and the Hazards of Modern Life*. Edited by Monica Caspter, 197–212. New York: Routledge, 2003.

BenDor, Jan. "Karman as 'Hero': Abortion Innovator Exhibits Sexism." *Herself* 1973, 1.

Benne, Kenneth. *A Conception of Authority*. New York: Russell and Russell, 1943.

———. "Democratic Ethics in Social Engineering." *Progressive Education* 26, no. 7 (1949): 201–7.

———. "History of the T Group in the Laboratory Setting." In *T-Group Theory and Laboratory Method: Innovation in Re-Education*. Edited by Leland Bradford, Jack Gibb, and Kenneth Benne. New York: John Wiley and Sons, 1964.

————. "Leaders Are Made, Not Born." *Childhood Education* 24, no. 5 (1948): 203–13.

Berg, Marc. "Turning a Practice into a Science: Reconceptualizing Postwar Medical Practice." *Social Studies of Science* 25, no. 3 (1995): 437–76.

Berg, Marc, and Stefan Timmermans. "Standardization in Action: Achieving Local Universality through Medical Protocols." *Social Studies of Science* 27 (1997): 273–305.

Bergeron, Suzanne. *Fragments of Development: Nation, Gender, and the Space of Modernity*. Ann Arbor: University of Michigan Press, 2004.

Bernays, Edward. *Propaganda*. New York: Horace Liveright, 1928.

Bernstein, Gerald. "The Los Angeles Experience." *Family Planning Perspectives* 2, no. 1 (1970): 39–41.

Biko, Steve. "The Definition of Black Consciousness." In *Philosophy from Africa: A Text with Readings*. Edited by P. H. Coetzee and A. P. Roux, 360–63. Halfway House: International Thompson Publishing, 1998.

Bishop, Amie, Jacqueline Sherris, Vivien Tsu, and M. Kilbourne-Brook. "Cervical Dysplasia Treatment: Key Issues for Developing Countries." *Bulletin of the Pan American Health Organization* 30, no. 4 (1996): 378–86.

Bishop, Amie, Elisa Wells, Jacqueline Sherris, Vivien Tsu, and Barbara Crook. "Cervical Cancer: Evolving Prevention Strategies for Developing Countries." *Reproductive Health Matters* 6, (November 1995): 60–71.

Bloom, Lisa. "Constructing Whiteness: *Popular Science* and the *National Geographic* in the Age of Multiculturalism." *Configurations* 2, no. 1 (1994): 15–32.

Bogdanich, Walt. "Lax Laboratories: The Pap Test Misses Much Cervical Cancer through Labs' Errors." *Wall Street Journal*, November 2, 1987.

Bonneuil, Christophe. "Development as Experiment: Science and State Building in Late Colonial and Postcolonial Africa, 1930–1970." *Osiris* 15 (2001): 258–81.

Bose, Sarmila. *Dead Reckoning: Memories of the 1971 Bangladesh War*. New York: Columbia University Press, 2011.

Boston Women's Health Book Collective. *Our Bodies, Ourselves: A Book by and for Women*. New York: Simon and Schuster, 1971.

Bouvier, Leon. "Doctors and Nurses: A Demographic Profile." Center for Immigration Study, www.cis.org. February 1998.

Branch, Marie, and Phyllis Paxton. *Providing Safe Nursing Care for Ethnic People of Color*. New York: Appleton-Century-Crofts, 1976.

Brandt, Allan. "Behavior, Disease, and Health in the Twentieth-Century United States: The Moral Valence of Individual Risk." In *Morality and Health*. Edited by Allan Brandt and Paul Rozin, 53–78. New York: Routledge.

Breslow, Lester. *History of Cancer Control in the U.S., 1946–1971*. Washington, D.C.: National Cancer Institute, 1978.

Briggs, Laura. "The Race of Hysteria: 'Overcivilization' and the 'Savage' Woman in Late Nineteenth-Century Obstetrics and Gynecology." *American Quarterly* 52, no. 2 (2000): 246–73.

————. *Reproducing Empire: Race, Sex, Science and U.S. Imperialism in Puerto Rico.* Berkeley: University of California Press, 2003.

Brody, Jane. "Physicians throughout the World Are Studying New, Simple Technique for Terminating Pregnancy." *New York Times.* December 20, 1973, 26.

Brown, Elsa Barkley. "Polyrhythms and Improvization: Lessons for Women's History." *History Workshop Journal* 31, no. 1 (1991): 85–90.

Brown, Phil, and Edwin Mikkelsen. *No Safe Place: Toxic Waste, Leukemia, and Community Action.* Berkeley: University of California Press, 1997.

Burke, Richard, and Warren Bennis. "Changes in Perception of Self and Others during Human Relations Training." *Human Relations* 14, no. 2 (1961): 165–82.

Butler, Edith. "The First National Conference on Black Women's Health Issues." In *Women's Health: Readings on Social, Economic, and Political Issues.* Edited by Nancy Worcester and Marianne Whatley, 37–14. Dubuque: Kendall, 1988.

Butler, Judith. *Bodies That Matter: On the Discursive Limits of "Sex."* New York: Routledge, 1993.

————. *Precarious Life: The Powers of Mourning and Violence.* New York: Verso, 2006.

Byrd, Michael, and Linda Clayton. *An American Health Dilemma: Race, Medicine, and Health Care in the United States, 1900–2000.* New York: Routledge, 2002.

Cambrosio, Alberto, Peter Keating, Thomas Schlich, and George Weisz. "Regulatory Objectivity and the Generation and Management of Evidence in Medicine." *Social Science and Medicine* 63, no. 1 (2006): 189–99.

Carpenter, V., and B. Colwell. "Cancer Knowledge, Self-Efficacy, and Cancer Screening Behaviors among Mexican-American Women." *Journal of Cancer Education* 10, no. 4 (1995): 217–22.

Carson, Ann, et al. *Choosing a Pap Smear Lab: A Guide for the Health Care Provider.* San Francisco: Coalition for the Medical Rights of Women, 1977.

Caruana, Stephanie. "The Complete Facts on Abortion." *Playgirl.* 1973, 80.

Casper, Monica, and Adele Clarke. "Making the Pap Smear into the 'Right Tool' for the Job: Cervical Cancer Screening in the USA, Circa 1940–95." *Social Studies of Science* 28, no. 2 (1998): 255–90.

Chatterjee, Nilanjana, and Nancy Riley. "Planning an Indian Modernity: The Gendered Politics of Fertility Control." *Signs* 26, no. 1 (2001): 811–45.

Chatterjee, Partha. *The Politics of the Governed: Reflections on Popular Politics in Most of the World.* New York: Columbia University Press, 2006.

Chaudhuri, Prasun. "Killer at Large." *Telegraph*, November 21, 2005.

Ch'eng, Pang-chung, Ch'in-jui Yang, and Kuo-ying Fang. "Use of the 'Fire Cup' Type Uterine Suction Bottle for Induced Abortion." *Chinese Journal of Nursing* 2, no. March (1966): 89–91.

Chesler, Ellen. *Woman of Valor: Margaret Sanger and the Birth Control Movement in America.* New York: Simon and Schuster, 1992.

Chico Feminist Women's Health Center. "Feminist Women's Health Center Chronology." Photocopy. Chico, Calif., 1994.

Clark, Sawin. "George Papanicolaou and the 'Pap' Test." *Endocrinologist* 12, no. 4 (2002): 267–72.

Clarke, Adele. *Disciplining Reproduction: Modernity, American Life Sciences and the "Problems of Sex."* Berkeley: University of California Press, 1998.

———. "From Grounded Theory to Situated Analysis: What's New? Why? How?" In *Developing Grounded Theory*. Edited by Jan Morse, et al. Walnut Creek, Calif.: Left Coast Press, 2009.

Clarke, Adele, and Monica Casper. "From Simple Technology to Complex Arena: Classification of Pap Smears, 1917–1990." *Medical Anthropology Quarterly* 10, no. 4 (1996): 601–23.

Clarke, Adele, et al. "Biomedicalization: Technoscientific Transformations of Health, Illness, and U.S. Biomedicine." *American Sociological Review* 68, no. 2 (2003): 161–94.

Clarke, Adele, and Theresa Montini. "The Many Faces of RU-486: Tales of Situated Knowledges and Technological Contestations." *Science, Technology and Human Values* 18, no. 1 (1993): 42–78.

Clarke, Adele, and Lisa Moore. "Clitoral Conventions and Transgressions: Graphic Representations in Anatomy Texts, C1900–1991." *Feminist Studies* 21, no. 2 (1995): 255–301.

Cleland, John, and W. Parker Mauldin. "The Promotion of Family Planning by Financial Payments: The Case of Bangladesh." *Studies in Family Planning* 22, no. 1 (1991): 1–18.

Cohen, Lawrence. "Operability, Bioavailability, and Exception." In *Global Assemblages: Technology, Politics, and Ethics as Anthropological Problems*. Edited by Aihwa Ong and Stephen Collier, 79–90. Malden, Mass.: Blackwell, 2005.

Collier, Steven. "Topologies of Power: Foucault's Analysis of Political Government beyond 'Governmentality.'" *Theory, Culture and Society* 26, no. 6 (2009): 78–108.

Collins, Patricia Hill. *Black Feminist Thought: Knowledge, Consciousness, and the Politics of Empowerment*. New York: Routledge, 1991.

Combahee River Collective. "Combahee River Collective Statement." In *Capitalist Patriarchy and the Case for Socialist Feminism*. Edited by Zillah Eisenstein, 247–55. New York: Monthly Review, 1978.

———. "Why Did They Die?" *Radical America* 13, no. 6 (1979): 41–50.

Composite Report of the President's Committee to Study the United States Military Assistance Program. Washington, D.C.: Government Printing Office, 1959.

Connelly, Matthew. *Fatal Misconception: The Struggle to Control World Population*. Cambridge: Harvard University Press, 2008.

Cooper, R. S., and R. David. "The Biological Concept of Race and Its Application to Public Health and Epidemiology." *Journal of Health Politics and Policy Law* 11 (1986): 97–116.

Correa, Sonia. *Population and Reproductive Rights: Feminist Perspectives from the South*. London: Zed Books, 1994.

Craeger, Angela, Elizabeth Lunbeck, and Norton Wise, eds. *Science without Laws: Model Systems, Cases, Exemplary Narratives*. Durham: Duke University Press, 2007.

Crawford, R. "Healthism and the Medicalization of Everyday Life." *International Journal of Health Services* 10 (1980): 365–88.

Crichtlow, Donald. *Intended Consequences: Birth Control, Abortion, and the Federal Government in Modern America*. Oxford: Oxford University Press, 1999.

Cruikshank, Barbara. *The Will to Empower: Democratic Citizens and Other Subjects*. Ithaca: Cornell University Press, 1999.

Dalton, Amy. "Rediscovering the Arusha Declaration and Abandoning the NGO Model." In *Indymedia*, at la.indymedia.org/news. Search by author. 2008.

Daston, Lorraine. "The Moral Economy of Science." *Osiris* 10 (1995): 2–24.

Daston, Lorraine, and Peter Galison. *Objectivity*. Cambridge: Zone Books, 2007.

Davidson, Arnold. "In Praise of Counter-Conduct." Paper presented at "Foucault across the Disciplines" conference, University of California, Santa Cruz, March 2008.

Davies, Carole Boyce. *Left of Karl Marx: The Political Life of Black Communist Claudia Jones*. Durham: Duke University Press, 2007.

Davis, Angela. "Sick and Tired of Being Sick and Tired: The Politics of the National Black Women's Health Project." In *The Black Women's Health Book*. Edited by Evelyn White, 18–26. Seattle: Seal, 1990.

Davis, Audrey. "Life Insurance and the Physical Examination: A Chapter in the Rise of American Medical Technology." *Bulletin of the History of Medicine* 55 (1981): 392–406.

Davis, Kathy. *The Making of Our Bodies, Ourselves: How Feminism Travels across Borders*. Durham: Duke University Press, 2007.

Davis-Floyd, Robbie. *Birth as an American Rite of Passage*. Berkeley: University of California Press, 1993.

D'Costa, Bina. *Nation-building, Gender and War Crimes in South Asia*. London: Routledge, 2011.

Deleuze, Gilles. *The Fold: Leibniz and the Baroque*. Translated by Tom Conley. Minneapolis: University of Minnesota Press, 1992.

Deleuze, Gilles, and Félix Guattari. *A Thousand Plateaus: Capitalism and Schizophrenia*. Translated by Brian Massumi. Minneapolis: University of Minnesota Press, 1987.

Deleuze, Gilles, and Claire Parnet. *Dialogues*. Translated by Hugh Tomlinson and Barbara Habberjam. New York: Columbia University Press, 1977.

Delgado, Richard, and Jean Stefancic, eds. *Critical White Studies: Looking beyond the Mirror*. Philadelphia: Temple University Press, 1997.

Deming, Barbara. *We Cannot Live without Our Lives*. New York: Grossman Publishers, 1974.

Denniston, R. W. "Cancer Knowledge, Attitudes, and Practices among Black Americans." In *Cancer among Black Populations*. Edited by C. Medlin and G. P. Murphy, 225–35. New York: Alan R Liss, 1981.

Devesa, S. S., D. T. Silverman, J. L. Young, et al. "Cancer Incidence and Mortality Trends among Whites in the United States, 1947–1984." *Journal of the National Cancer Institute* 79 (1987): 701.

Dichter, Ernest. *The Strategy of Desire*. New York: Doubleday, 1960.

Dicochea, Perlita. "Chicana Critical Rhetoric: Recrafting La Causa in Chicana Movement Discourse, 1970–1979." *Frontiers: A Journal of Women's Studies* 25, no. 1 (2004): 77–92.

Dixon-Mueller, Ruth. "Innovations in Reproductive Health Care: Menstrual Regulation Policies and Programs in Bangladesh." *Studies in Family Planning* 19, no. 3 (1988): 129–40.

Downer, Carol. "Self Help: What Is It?" Mimeograph. Los Angeles, 1973.

————. "What Makes the Feminist Women's Health Center 'Feminist'?" *Off Our Backs*. June (1974): 2–5.

Downer, Carol, Lorraine Rothman, and Eleanor Snow, "F.W.H.C. Response." *Off Our Backs*. August/September (1974): 17.

Dreifus, Claudia. Introduction to *Seizing Our Bodies*. Edited by Claudia Dreifus. New York: Vintage, 1977.

Dryfoos, Joy. "Planning Family Planning in Eight Metropolitan Areas." *Family Planning Perspectives* 3, no. 1 (1971): 11–15.

Du Bois, W. E. B. *The Souls of Black Folks*. Chicago: A. C. McClurg and Co, 1903.

Duster, Troy. "Feedback Loops in the Politics of Knowledge Production." In *The Governance of Knowledge*. Edited by Nico Stehr, 139–60. New Brunswick, N.J.: Transaction Books, 2003.

Dyer, Richard. "White." *Screen* 29, no. 4 (1998): 44–65.

Echols, Alice. *Daring to Be Bad: Radical Feminism in America, 1967–75*. Minneapolis: University of Minnesota Press, 1989.

Edwards, Hodee. "Housework and Exploitation: A Marxist Analysis." *No More Fun and Games: A Journal of Female Liberation*, no. 4 (1971): 92–100.

Ehrenreich, Barbara, and John Ehrenreich. *The American Health Empire: Power, Profits, and Politics*. New York: HealthPAC Book, 1971.

Enloe, Cynthia. *Maneuvers: The International Politics of Militarizing Women's Lives*. Berkeley: University of California Press, 2004.

Epstein, Steven. *Inclusion: The Politics of Difference in Medical Research*. Chicago: University of Chicago Press, 2007.

Evans, Sara. *Personal Politics: The Roots of Women's Liberation in the Civil Rights Movement and the New Left*. New York: Vintage, 1979.

Fairchild, Amy. *Science at the Borders: Immigrant Medical Inspection and the Shaping of the Modern Industrial Labor Force*. Baltimore: Johns Hopkins University Press, 2003.

Fanon, Franz. *Black Skin, White Masks*. Translated by Constance Farrington. New York: Grove Press, 1994.

————. "Medicine and Colonialism." In *A Dying Colonialism*, 121–46. New York: Groves Press, 1965.

Farmer, Paul. *Aids and Accusation: Haiti and the Geography of Blame*. Berkeley: University of California Press, 1992.

Federation of Feminist Women's Health Centers (FFWHC). *How to Stay out of the Gynecologist's Office*. Hollywood: Women to Women Publications, 1981.

————. *A New View of a Woman's Body*. Los Angeles: Feminist Health Press, 1991.

Federici, Silvia. *Wages against Housework*. Bristol: Falling Wall Press, 1973.

Feldstein, Ruth. *Motherhood in Black and White: Race and Sex in American Liberalism, 1930–1965*. Ithaca: Cornell University Press, 2000.

Feminist Women's Health Center (FWHC). *Abortion in a Clinical Setting*. Los Angeles: Feminist Women's Health Center, 1974.

————. *Self-Help Study: Observing Changes in the Menstrual Cycle*. Mimeograph. Los Angeles: Feminist Women's Health Center, 1975.

————. *Well Woman Health Care in Woman Controlled Clinics*. Los Angeles: Feminist Women's Health Center, 1976.

Ferguson, Roderick. *Aberrations in Black: Toward a Queer of Color Critique*. Minneapolis: University of Minnesota, 2004.

————. "Of Our Normative Strivings: African American Studies and the Histories of Sexuality." *Social Text* 23, nos. 3–4 (2005): 85–100.

Fidler, K., D. Boyes, and A. J. Worth. "Cervical Cancer Detection in British Columbia: A Progress Report." *Journal of Obstetrics and Gynecology of British Columbia* 75 (1968): 392–404.

FINRRAGE-UBINIG International Conference 1989. "Declaration of Comilla." *Journal of Issues in Reproductive and Genetic Engineering* 4, no. 1 (1991): 73–74.

Firestone, Shulamith. *The Dialectic of Sex: The Case for Feminist Revolution*. New York: Morrow, 1970.

Foltz, Anne Marie, and Jennifer Kelsey. "The Annual Pap Test: A Dubious Policy Success." *Milbank Memorial Fund Quarterly* 56, no. 4 (1978): 426–62.

Forbes Group. "Plotting the Future of Cytotechnology: An Environmental Analysis of the Driving Forces of Cytology." American Society of Cytopathology, 2007.

Formisano, Ronald. *Boston against Busing: Race, Class, and Ethnicity in the 1960s and 1970s*. Chapel Hill: University of North Carolina Press, 2003.

Foucault, Michel. *The Birth of Biopolitics: Lectures at the College De France, 1978–79*. Translated by Graham Burchell. Edited by Michel Senellart et al. New York: Palgrave Macmillan, 2008.

————. *History of Sexuality*. Vol. 1: *An Introduction*. Translated by Robert Hurley. New York: Random House, 1978.

————. "Le sujet et le pouvoir." In *Dits et écrits IV, 1980–1988*. Edited by F. Ewald and J. Lagrange, 222–43. Paris: Gallimard, 1994.

————. *Security, Territory, Population: Lectures at the College De France, 1977–1978*.

Translated by Graham Burchell. Edited by Michel Senellart et al. New York: Palgrave Macmillan, 2007.

———. *Society Must Be Defended: Lectures at the Colleges De France, 1975–76*. Translated by David Macey. New York: Picador, 2003.

———. "The Subject and Power." In *Michel Foucault—Power: Essential Works of Foucault 1954–1984*. Edited by James Faubion, 326–48. New York: New Press, 2000.

Franklin, Sarah. *Dolly Mixtures: The Remaking of Genealogy*. Durham: Duke University Press, 2007.

Friedan, Betty. *The Feminine Mystique*. New York: Dell, 1963.

Gage, J. C., C. Ferreccio, and M. Gonzales. "Follow-up Care of Women with Abnormal Cytology in a Low-Resource Setting." *Cancer Detection and Prevention* 27 (2007): 466–71.

Gage, Suzanne, ed. *Lesbian Health Activism: The First Wave. Feminist Writings from the Early Lesbian Health Movement*. Los Angeles: Feminist Health Press, 1973.

Galloway, Alexander. *Protocol: How Control Exists after Decentralization*. Cambridge: MIT Press, 2006.

Gardner, Kirsten. *Early Detection: Women, Cancer, and Awareness Campaigns in the Twentieth-Century United States*. Chapel Hill: University of North Carolina Press, 2006.

Gerlach, Luther P., and Virginia H. Hine. *People, Power, Change: Movements of Social Transformation*. Indianapolis: Bobbs-Merrill, 1970.

Germain, Adrienne, et al., eds. *Reproductive Tract Infections: Global Impact and Priorities for Women's Reproductive Health*. New York: Plenum Press, 1992.

Germain, Adrienne, and Jane Ordway. *Population Control and Women's Health: Balancing the Scales*. New York: IWHC, 1989.

Gilroy, Paul. "One Nation under a Groove: The Cultural Politics of 'Race' and Racism in Britain." In *Anatomy of Racism*. Edited by David Theo Goldberg, 163–82. Minneapolis: University of Minnesota Press, 1990.

Gluck, Sherna Berger, et al. "Whose Feminism, Whose History? Reflections on Excavating the History of (the) US Women's Movement." In *Community Activism and Feminist Politics: Organizing across Race, Class, and Gender*. Edited by Nancy Naples, 57–80. New York: Routledge, 1998.

Goldberg, David Theo. *The Racial State*. Malden, Mass.: Blackwell, 2002.

———. *Racial Subjects: Writing on Race in America*. New York: Routledge, 1997.

Goldsmith, Philip. "Interview with Harvey Karman." *Journal of Environmental Quality* 2 (1973): 48.

Gordon, Colin. "Government Rationality: An Introduction." In *The Foucault Effect: Studies in Governmentality*. Edited by Graham Burchelle, Colin Gordon, and Peter Miller, 1–51. Chicago: University of Chicago Press, 1991.

Governor's Commission on the Los Angeles Riots. "Violence in the City—An End or a Beginning?" Los Angeles, 1965.

Grayson, Deborah. "'Necessity Was the Midwife of Our Politics': Black Women's Health Activism in the 'Post'-Civil Rights Era (1980-1996)." In *Still Lifting, Still Climbing: African American Women's Contemporary Activism.* Edited by Kimberly Springer, 131-48. New York: New York University Press, 1999.

Greco, Monica. "The Politics of Indeterminacy and the Right to Health." *Theory, Culture and Society* 21, no. 6 (2004): 1-22.

Greene, Jeremy. *Prescribing by Numbers: Drugs and the Definition of Disease.* Baltimore: Johns Hopkins University Press, 2006.

Greenhalgh, Susan, and Edwin Winckler. *Governing China's Population: From Leninist to Neoliberal Biopolitics.* Stanford: Stanford University Press, 2005.

Gregg, Jessica. *Virtually Virgins: Sexual Strategies and Cervical Cancer in Recife Brazil.* Stanford: Stanford University Press, 2003.

Gregg, Jessica, and Robert Curry. "Explanatory Models for Cancer among African-American Women at Two Atlanta Health Centers: The Implications for a Cancer Screening Program." *Social Science and Medicine* 39, no. 4 (1994): 519-26.

Grewal, Inderpal, and Caren Kaplan, eds. *Scattered Hegemonies: Postmodernity and Transnational Feminist Practices.* Minneapolis: University of Minnesota Press, 1994.

Grossman, Rachel. "Woman's Place in the Integrated Circuit." *Radical America* 14, no. 1 (1980): 29-50.

"Guard Yourself" commercial. At www.youtube.com. May 2008.

Guevara, Che. "On Revolutionary Medicine." In *Che Guevara: Radical Writings on Guerilla Warfare, Politics and Revolution,* 13-25. Minneapolis: Filiquarian Publishing, 2006.

Gumbs, Alexis Pauline. "We Can Learn to Mother Ourselves: The Queer Survival of Black Feminism 1968-1996." PhD dissertation, Duke University, 2010.

Gutierrez, Elena. "Policing 'Pregnancy Pilgrims': Situating the Sterilization Abuse of Mexican-Origin Women in Los Angeles County." In *Women, Health, and Nation: Canada and the United States since 1945.* Edited by Georgina Feldberg et al. Montreal and Kingston: McGill-Queen's University Press, 2003.

———. *Fertile Matters: The Politics of Mexican-Origin Women's Reproduction.* Austin: University of Texas Press, 2011.

Halfon, Saul. *The Cairo Consensus: Demographic Surveys, Women's Empowerment and Regime Change in Population Policy.* Lanham, Md.: Lexington Books, 2006.

Hammonds, Evelynn. "Missing Persons: African American Women, Aids and the History of Disease." *Radical American* 20, no. 2 (1990): 7-23.

Hanish, Carol. "The Personal Is Political." In *Feminist Revolution.* Edited by Redstockings, 204-5. New York: Random House, 1975.

Haraway, Donna. "A Cyborg Manifesto: Science, Technology, and Socialist-Feminism in the Late Twentieth Century." In *Simians, Cyborgs, and Women,* 149-81. New York: Routledge, 1991.

———. "Manifesto for Cyborgs: Science, Technology and Socialist Feminism in the 1980s." *Socialist Review* 80 (1985): 65–108.

———. *Modest_Witness@Second_Millennium: Femaleman_Meets_OncoMouse: Feminism and Technoscience*. New York: Routledge, 1997.

———. "Situated Knowledges: The Science Question in Feminism and the Privilege of Partial Perspective." In *Simians, Cyborgs, and Women*, 183–202. New York: Routledge, 1991.

———. "The Virtual Speculum for a New World Order." In *Revisioning Women, Health and Healing: Feminist, Cultural and Technoscience Perspectives*. Edited by Adele Clarke and Virginia Olesen, 49–96. New York: Routledge, 1998.

Harden, Patricia, et al. "The Sisters Reply." In *Poor Black Women*. Edited by Robinson Patricia. Boston: New England Free Press, n.d.

Harding, Sandra. "'Strong Objectivity' and Socially Situated Knowledge." In *Whose Science? Whose Knowledge?*, 138–63. Ithaca: Cornell University Press, 1991.

———. "Subjectivity, Experience, and Knowledge: An Epistemology from/for the Rainbow Coalition Politics." In *Who Can Speak? Authority and Critical Identity*. Edited by Judith Roof and Robyn Wiegman, 120–36. Urbana: University of Illinois Press, 1995.

Hardt, Michael, and Antonio Negri. *Empire*. Cambridge: Harvard University Press, 2000.

———. *Multitude: War and Democracy in the Age of Empire*. New York: Penguin, 2004.

Harris, Duchess. "From the Kennedy Commission to the Combahee Collective: Black Feminist Organizing, 1960–80." In *Sisters in the Struggle: African American Women in the Civil Rights-Black Power Movement*. Edited by Bettye Collier-Thomas and V. P. Franklin, 280–305. New York: NYU Press, 2001.

Harstock, Nancy. "The Feminist Standpoint: Developing the Ground for a Specifically Feminist Historical Materialism." In *Discovering Reality: Feminist Perspectives on Epistemology, Metaphysics, Methodology and Philosophy of Science*. Edited by Sandra Harding and Merrill Hintikka, 283–310. Dordrecht: Reidel, 1983.

Hartmann, Betsy. *Reproductive Rights and Reproductive Wrongs: The Global Politics of Population Control*. Cambridge, Mass.: South End Press, 1995.

Hartmann, Betsy, and Hilary Standing. *Food, Saris and Sterilization: Population Control in Bangladesh*. Birmingham: Third World Publications, 1985.

Hecht, Gabrielle. *The Radiance of France: Nuclear Power and National Identity after World War II*. Cambridge: MIT Press, 1998.

Herman, Ellen. *The Romance of American Psychology: Political Culture in the Age of Experts*. Berkeley: University of California Press, 1997.

Herrington, Simon. "Do HPV-Negative Cervical Carcinomas Exist?—Revisited." *Journal of Pathology* 189, no. 1 (1999): 1–3.

Hinchliffe, Steve. "Scalography and Worldly Assemblies." Paper presented at Scalography Conference, Oxford. 2009.

Hinton, William. *Fanshen: A Documentary of Revolution in a Chinese Village*. New York: Monthly Review Press, 1967.

Hirsch, Lolly. "Second Women-Controlled Women's Health Center Conference," *Monthly Extract* 3, no. 2 (1974): 2.

———. Untitled, *Monthly Extract* 3, no. 2 (1974): 8.

———. "Practicing Health without a License," *Monthly Extract* (July/August 1977): 3.

Hochschild, Arlie. *The Managed Heart: Commercialization of Human Feeling*. Berkeley: University of California Press, 1983.

Hodgson, Dennis, and Susan Watkins. "Feminists and Neo-Malthusians: Past and Present Alliances." *Population and Development Review* 23, no. 3 (1997): 469–523.

Hoffman, Fredrick. *Cancer and Diet, with Facts and Observations of Related Subjects*. Baltimore: Williams and Wilkins, 1937.

———. "The Cancer Problems and Research, Delivered before the Canadian Public Health Association," 12–24. Montreal, 1925.

Hohmeister, Frederick. "The Gynecologic Examination." *Clinical Obstetrics and Gynecology* (1959): 1173–95.

Holden, Kerry, and David Demeritt. "Democratising Science? The Politics of Promoting Biomedicine in Singapore's Developmental State." *Environment and Planning D* 26, no. 1 (2008): 68–86.

Hong, Grace Kyungwon. *The Ruptures of American Capital: Women of Color Feminism and the Culture of Immigrant Labor*. Minneapolis: Minnesota, 2006.

Hornstein, Frances. *Lesbian Health Care*. Los Angeles: Feminist Women's Health Centers, 1973.

Hornstein, Frances, Carol Downer, and Shelly Farber. *Self Help and Health: A Report*. Washington Women's Self Help and LA Feminist Women's Health Center, 1976.

Hulbert, Randall, and Robert Settlage. "Birth Control and the Private Physician: The View from Los Angeles." *Family Planning Perspectives* 6, no. 1 (1974): 50–55.

Hulka, Barbara. "Motivation Technics in a Cancer Detection Program: Utilization of Community Resources." *American Journal of Public Health* 57, no. 2 (1967): 229–41.

Hull, Gloria, Patricia Bell-Scott, and Smith Barbara, eds. *All the Women Are White, All the Blacks Are Men, but Some of Us Are Brave: Black Women's Studies*. Boston: Feminist Press, 1982.

Hutchinson, M., and J. Morales. "Population-Based Study of Human Papillomavirus Infection and Cervical Neopolasia in Rural Costa Rica." *Journal of the National Cancer Institute* 92, no. 464–74 (2000).

IWHC. *Twenty Years, One Goal: IWHC's 20th Anniversary Report*. New York: IWHC, 2004.

IWHC, and WAND. *Challenging the Culture of Silence: Building Alliances to End Reproductive Tract Infections*. New York: International Women's Health Coalition, 1994.

Jagger, Alison. "Love and Knowledge: Emotion in Feminist Epistemology." In *Gender/Body/Knowledge*. Edited by Alison Jagger and Susan Bordo, 145–71. New Brunswick: Rutgers University Press, 1989.

Jain, Sarah Lochlann. *Injury: The Politics of Product Design and Safety Law in the United States*. Princeton: Princeton University Press, 2006a.

———. "Sentience and Slaver: The Struggle over the Short-Handled Hoe." In *Injury: The Politics of Product Design and Safety Law in the United States*. Princeton: Princeton University Press, 2006b.

James, Selma, and Mariarosa Dalla Costa. *The Power of Women and the Subversion of the Community*. Bristol: Falling Wall Press, 1973.

John, Mary. *Discrepant Dislocations: Feminism, Theory and Postcolonial Histories*. Berkeley: University of California Press, 1996.

Johnson, Lyndon B. "Address in San Francisco at the 20th Anniversary Commemorative Session of the United Nations." June 25, 1965. www.presidency.ucsb.edu.

Jordanova, Ludmilla. "Interrogating the Concept of Reproduction in the Eighteenth Century." In *Conceiving the New World Order: The Global Politics of Reproduction*. Edited by Faye Ginsburg and Rayna Rapp, 369–86. Berkeley: University of California Press, 1995.

———. *Sexual Visions: Images of Gender in Science and Medicine between the Eighteenth and Twentieth Centuries*. Madison: University of Wisconsin Press, 1989.

Kainyah, Rosalind. "Remarks." De Beers Group. 2007. At www.iwhc.org.

Kaplan, Laura. *The Story of Jane: The Legendary Underground Feminist Abortion Service*. Chicago: University of Chicago Press, 1997.

Karman, Harvey. "The Paramedic Abortionist." In *Use of Allied Health Personnel in Obstetrics and Gynecology*. Edited by John Marshall, 379–87. Hagerstown, Md: Harper & Row Publishers, 1972.

Karman, Harvey, and Richard Scotti. "Menstrual Regulation and Early Pregnancy Termination Performed by Paraprofessionals under Medical Supervision." *Contraception* 14, no. 4 (1976): 367–74.

Katz, Cindi. *Growing up Global: Economic Restructuring and Children's Everyday Lives*. Minneapolis: University of Minnesota, 2004.

———. "On the Grounds of Globalization: A Topography for Feminist Political Engagement." *Signs* 26, no. 4 (2001): 1213–34.

Kay, Bonnie, Adrienne Germain, and Maggie Bangser. *The Bangladesh Women's Health Coalition, Quality/Calidad/Qualité*. New York: Population Council, 1991.

Keller, Evelyn Fox. *A Feeling for the Organism: The Life and Work of Barbara McClintock*. New York: W. H. Freeman and Company, 1983.

Kelty, Christopher. *Two Bits: The Cultural Significance of Free Software*. Durham: Duke University Press, 2008.

"Killing of Babies Feared in Bengal." *New York Times*, March 8, 1972, 8.

Kirchhoff, Helen, and R. H. Rigdon. "Frequency of Cancer in the White and Negro: A Study Based upon Necropsies." *Southern Medical Journal* 49, no. 8 (1956): 834–41.

Klawiter, Maren. "Racing for the Cure, Walking Women, and Toxic Touring: Mapping Cultures of Action within the Bay Area Terrain of Breast Cancer." *Social Problems* 46 (1999): 104–26.

Kline, Wendy. *Bodies of Knowledge: Sexuality, Reproduction, and Women's Health in the Second Wave*. Chicago: University of Chicago Press, 2010.

Koblin, Helen. "Vaginal Politics, Dr. Karman, Abortionist." *Los Angeles Free Press*, June 9, 1972.

Koss, Leopold. "The Attack on the Annual 'Pap Smear.'" *Acta Cytological: The Journal of Clinical Cytology* 24, no. 3 (1980): 181–83.

Krieger, Nancy. "Racism, Sexism, and Social Class: Implications for Studies of Health, Disease, and Well-Being." *American Journal of Preventative Medicine* 9, no. 2 (1993): 82–122.

———. "Refiguring 'Race': Epidemiological, Racialized Biology, and Biological Expressions of Race Relations." *International Journal of Health Services* 30, no. 1 (2000): 211–16.

Kripal, Jeffrey. *Esalen: America and the Religion of No Religion*. Chicago: University of Chicago Press, 2007.

Landecker, Hannah. *Culturing Life: When Cells Became Technologies*. Cambridge: Harvard University Press, 2007.

Landman, Lynn. "Los Angeles' Experiment with Functional Coordination: A Progress Report." *Family Planning Perspectives* 3, no. 2 (1971): 4–15.

Lang, Sabine. "The NGOization of Feminism." In *Transitions, Environments, Translations: Feminisms in International Politics*. Edited by Joan W. Scott, Caren Kaplan, and Debra Keates, 101–20. New York: Routledge, 1997.

Lannin, D. R., et al. "Influence of Socioeconomic and Cultural Factors on Racial Differences in Late-Stage Presentation of Breast Cancer." *JAMA* 279, no. 22 (1998): 1801–7.

Lasch-Quinn, Elisabeth. *Race Experts: How Racial Etiquette, Sensitivity Training, and New Age Therapy Hijacked the Civil Rights Revolution*. New York: W. W. Norton, 2001.

Latham, Melanie. *Regulating Reproduction: A Century of Conflict in Britain and France*. Manchester: Manchester University Press, 2002.

Latour, Bruno. *Science in Action*. Cambridge: Harvard University Press, 1987.

Laufe, Leonard. "Menstrual Regulation: International Perspectives." In *Pregnancy Termination. PARFR Series on Fertility Regulation*. Edited by G. I. Zatuchni, J. J. Sciarra, and J. J. Speidel, 78–81. Hagerstown, Md.: Harper and Row, 1979.

Lazcano-Ponce, Eduardo César, et al. "Cervical Cancer Screening in Developing Countries: Why Is It Ineffective? The Case of Mexico." *Archives of Medical Research* 30, no. 3 (1999): 240–50.

Lazzarato, Maurizio. "Immaterial Labor." In *Radical Thought in Italy: A Potential Politics*. Edited by Michael Hardt and Paolo Virno, 133–47. Minneapolis: University of Minnesota Press, 1996.

Leach, William. *Land of Desire: Merchants, Power and the Rise of a New American Culture*. New York: Pantheon Books, 1993.

Lee, Laura Kim. "Changing Selves, Changing Society: Human Relations Experts and the Invention of T Groups, Sensitivity Training, and Encounter in the United States, 1938–1980." PhD dissertation, UCLA, 2002.

Lerner, Barron. "Beyond Informed Consent: Did Cancer Patients Challenge Their Physicians in the Post-World War II Era?" *Journal of the History of Medicine and Allied Sciences* 59, no. 4 (2004): 507–21.

———. *The Breast Cancer Wars: Hope, Fear, and the Pursuit of a Cure in Twentieth-Century America*. Oxford: Oxford University Press, 2001.

Leste, Judy, et al. "Calling It Quits." *Off Our Backs* June (1974): 2–5.

Lewin, Kurt, and Ronald Lippitt. "An Experimental Approach to the Study of Autocracy." *Sociometry* 1, nos. 3/4 (1938): 292–300.

Lewis, Andrea, and Beverly Smith. "Looking at the Total Picture: A Conversation with Health Activist Beverly Smith." In *The Black Women's Health Book: Speaking for Ourselves*. Edited by Evelyn White, 172–81. Seattle: Seal, 1990.

Lipsitz, George. *The Possessive Investment in Whiteness: How White People Benefit from Identity Politics*. Philadelphia: Temple University Press, 1998.

Lorde, Audre. *A Burst of Light*. Ithaca: Firebrand Books, 1988.

Lowe, Lisa. "The Worldliness of Intimacy." In *Edward Said: The Legacy of a Public Intellectual*. Edited by Debjani Ganuly and Ned Curthoys, 121–51. Melbourne: Melbourne University Press, 2007.

Luibheid, Eithne. *Entry Denied: Controlling Sexuality at the Border*. Minneapolis: University of Minnesota Press, 2002.

Luker, Kristin. *Abortion and the Politics of Motherhood*. Berkeley: University of California Press, 1984.

MacPherson, C. B. *The Political Theory of Possessive Individualism: Hobbes to Locke*. Oxford: Clarendon Press, 1962.

Maginnis, Pat, and Lena Phelan. *The Abortion Handbook for Responsible Women*. North Hollywood: Contact Books, 1969.

Mahoney, Martha. "Residential Segregation and White Privilege." In *Critical White Studies: Looking beyond the Mirror*. Edited by Richard Delgado and Jean Stefancic, 273–76. Philadelphia: Temple University Press, 1997.

"Making Foreign Aid Work." *New York Times*, July 26, 1959.

Marston, Sallie. "The Social Construction of Scale." *Progress in Human Geography* 24, no. 2 (2000): 219–42.

Martin, Emily. *The Woman in the Body*. Boston: Beacon Press, 1989.

Martin, R. E. "A Review of a Prison Cervical Cancer Screening Program in British Columbia." *Canadian Journal of Public Health* 89, no. 6 (1998): 382–86.

Marton, Kati. "Remarks by IWHC board chair, New York." January 19, 2006. In IWHC, www.iwhc.org.

Marx, J. L. "The Annual Pap Smear: An Idea Whose Time Has Gone?" *Science* 205 (1979): 177–78.

Maslow, Abraham. *Motivation and Personality*. New York: Harper, 1954.

———. "Self-Actualizing People: A Study of Psychological Health." In *The Self: Ex-*

plorations in Personal Growth. Edited by C. E. Moustakas. New York: Harper and Row, 1956.

Massumi, Brian. *Parables for the Virtual: Movement, Affect, Sensation*. Durham: Duke University Press, 2002.

Matthews, Nancy. *Confronting Rape: The Feminist Anti-Rape Movement and the State*. New York: Routledge, 1994.

Mazur, Paul. *American Prosperity*. New York: Doubleday, 1928.

Mbembe, Achille. "Necropolitics." *Public Culture* 15, no. 1 (2003): 11–40.

McGregor, Deborah Kuhn. *From Midwives to Medicine: The Birth of American Gynecology*. New Brunswick, N.J.: Rutgers University Press, 1998.

McNamara, Mary. "Pap Smears: Testing the Tests." *Ms*, April 1988, 65–67.

Memmi, Albert. *The Colonizer and the Colonized*. Translated by Howard Greenfeld. Boston: Beacon Press, 1965.

Menen, Audrey. "The Rapes of Bangladesh: The Rapes." *New York Times*, July 23, 1972, 10–15.

Mitchell, Timothy. "Fixing the Economy." *Cultural Studies* 12, no. 1 (1998): 82–101.

Mitchinson, Wendy. *Giving Birth in Canada, 1900–1950*. Toronto: University of Toronto Press, 2002.

Mol, Annemarie. *The Body Multiple: Ontology in Medical Practice*. Durham: Duke University Press, 2002.

Mol, Annemarie, and John Law, eds. *Complexities: Social Studies of Knowledge Practices*. Durham: Duke University Press, 2002.

Molina, Natalia. *Fit to Be Citizens? Public Health and Race in Los Angeles, 1879–1939*. Berkeley: University of California Press, 2006.

Montreal Health Press. *The Birth Control Handbook*. Montreal: Montreal Health Press. 1968.

Mookherjee, Nayanika. "Available Motherhood: Legal Technologies, 'States of Exception' and the Dekinning of 'War-Babies' in Bangladesh." *Childhood* 14, no. 3 (2007): 339–54.

———. "Gendered Embodiments: Mapping the Body-Politic of the Raped Woman and the Nation in Bangladesh." In *Critical Reflections on Gender and the South Asian Diaspora*. Edited by Nirmal Puwar and Parvati Raghuram, 157–77. Oxford: Berg, 2003.

———. *The Spectral Wound: Sexual Violence, Public Memories and the Bangladesh War of 1971*. Durham: Duke University Press, forthcoming.

Moore, Kelly. *Disrupting Science: Social Movements, American Scientists, and the Politics of the Military, 1945–1975*. Princeton: Princeton University Press, 2008.

Moraga, Cherrie, and Gloria Anzaldúa, eds. *This Bridge Called My Back: Writings by Radical Women of Color*. New York: Kitchen Table, Women of Color Press, 1981.

Mordini, Emilio. "Biowarfare as a Biopolitical Icon." *Poiesis and Praxis* 3, no. 4 (2005): 242–55.

Morgan, Kathryn. "Contested Bodies, Contested Knowledges: Women, Health and the Politics of Medicalization." In *The Politics of Women's Health: Exploring Agency and Autonomy*. Edited by Susan Sherwin, 83–121. Philadelphia: Temple University Press, 1998.

Morgan, Robin. Introduction to *Circle One: A Woman's Beginning Guide to Self Health and Sexuality*. Edited by Elizabeth Campbell and Vicki Ziegler. Colorado Springs: Circle One, 1975.

Morgen, Sandra. *Into Our Own Hands: The Women's Health Movement in the United States*. New Brunswick: Rutgers University Press, 2002.

Mueller-Wille, Staffan. "Nature as a Marketplace: The Political Economy of Linnaean Botany." *History of Political Economy* 35 (2003): 154–72.

Mueller-Wille, Staffan, and Hans-Jörg Rheinberger, eds. *Heredity Produced: At the Crossroads of Biology, Politics, and Culture, 1500–1870*. Cambridge: MIT Press, 2007.

Muñoz, José Esteban. *Disidentifications: Queers of Color and the Performance of Politics*. Minneapolis: University of Minnesota Press, 1999.

Murphy, Michelle. "Distributed Reproduction." In *Corpus: An Interdisciplinary Reader on Bodies and Knowledge*. Edited by Monica Casper and Paisley Currah. New York: Palgrave, 2011.

———. "Economization of Life: Calculative Infrastructures of Population and Economy." In *Relational Ecologies: Subjectivity, Sex, Nature and Architecture*. Edited by Peg Rawes. London: Routledge, forthcoming.

———. "Immodest Witnessing: The Epistemology of Vaginal Self-Examination in the U.S. Feminist Self-Help Movement." *Feminist Studies* 30, no. 1 (2004): 115–47.

———. "Liberation through Control in the Body Politics of U.S. Radical Feminism." In *The Moral Authority of Nature*. Edited by Lorraine Daston and Fernando Vidal, 331–55. Chicago: University of Chicago Press, 2003.

———. "Sex and Gender." In *The Palgrave Dictionary of Transnational History*. Edited by Akira Iriye and Pierre-Yves Saunier. London: Palgrave, 2008.

———. *Sick Building Syndrome and the Problem of Uncertainty: Environmental Politics, Technoscience, and Women Workers*. Durham: Duke University Press, 2006.

———. "Uncertain Exposures and the Privilege of Imperception: Activist Scientists and Race at the U.S. Environmental Protection Agency." In *Landscapes of Exposure: Knowledge and Illness in Modern Environments*. *Osiris* 19. Edited by Gregg Mitman, Michelle Murphy, and Christopher Sellers, 266–82. Chicago: University of Chicago Press, 2004.

Myers, Natasha. "Molecular Embodiments and the Body-Work of Modeling in Protein Crystallography." *Social Studies of Science* 38, no. 2 (2008): 163–99.

Nash, Linda. *Inescapable Ecologies: A History of Environment, Disease, and Knowledge*. Berkeley: University of California Press, 2006.

Negri, Antonio. "Value and Affect." *boundary 2* 26, no. 2 (1999): 77–88.

Nelson, Alondra. *Body and Soul: The Black Panther Party and the Fight against Medical Discrimination*. Minneapolis: University of Minnesota Press, 2011.

Nelson, Jennifer. "'Hold Your Head up and Stick Our Your Chin': Community Health and Women's Health in Mound Bayou, Mississippi." *NWSA Journal* 17, no. 1 (2005): 99–118.

———. *Women of Color and the Reproductive Rights Movement.* New York: New York University Press, 2003.

Nelson, Theodor. *Computer Lib: You Can and Must Understand Computers Now / Dream Machines: New Freedoms through Computer Screens—A Minority Report.* South Bend, Ind.: by the author, 1974.

Nettleton, Sarah. *Power, Pain, and Dentistry.* Philadelphia: Open University Press, 1992.

Newman, Louise Michele. *White Women's Rights: The Racial Origins of Feminism in the United States.* Oxford: Oxford University Press, 1999.

Nguyen, Vinh-Kim. "Antiretroviral Globalism, Biopolitics, and Therapeutic Citizenship." In *Global Assemblages: Technology, Politics, and Ethics as Anthropology.* Edited by Aihwa Ong and Stephen Collier, 124–44. Oxford: Blackwell, 2005.

———. "Government- by-Exception: Enrolment and Experimentality in HIV Treatment Programmes in Africa." *Social Theory & Health* 7, no. 3 (2009): 196–217.

Nonnenmacher, B., N. Hubbert, and R. Kirnbauer. "Serologic Response to Human Papillomavirus Type 16 Virus-Like Particles in HPV-16 DNA-Positive Invasive Cervical Cancer and Cervical Intraepithelial Neoplasia Grade III Patients and Controls from Colombia and Spain." *Journal of Infectious Disease* 172, no. 1 (1995): 19–25.

Norman, Brian. "The Consciousness-Raising Document, Feminist Anthologies and Black Women in Sisterhood Is Powerful." *Frontiers: A Journal of Women's Studies* 26, no. 3 (2005): 38–64.

———. "'We' in Redux: The Combahee River Collective's 'A Black Feminist Statement.'" *Differences* 18, no. 2 (2007): 103–32.

O'Connor, Alice. *Poverty Knowledge: Social Science, Social Policy, and the Poor in Twentieth-Century U.S. History.* Princeton: Princeton University Press, 2002.

Ogbar, Jeffrey. *Black Power: Radical Politics and African American Identity.* Baltimore: Johns Hopkins Press, 2005.

Orleans, Leo. "Abortion." *Studies in Family Planning* 4, no. 8 (1973): 198–202.

Ortiz, Ana Teresa. "'Bare-Handed' Medicine and Its Elusive Patients: The Unstable Construction of Pregnant Women and Fetuses in Dominican Obstetrics Discourse." *Feminist Studies* 23, no. 2 (1997): 263–88.

Oudshoorn, Nelly. "The Decline of the One-Size-Fits-All Paradigm, or, How Reproductive Scientists Try to Cope with Postmodernity." In *Between Monsters, Goddesses and Cyborgs: Feminist Confrontations with Science, Medicine and Cyberspace.* Edited by Nina Lykke and Rosi Braidotti, 153–72. London: Zed, 1996.

P'an, Jun-min. "Application of the Simple Negative Pressure Bottle and Its Removable Hand-Operated Suction Equipment." *Chinese Journal of Obstetrics and Gynecology* 11, no. 6 (1965): 401–3.

Papanicolaou, George. "New Cancer Diagnosis." Paper presented at the Proceedings of the Third Race Betterment Conference. 1928.

———. "The Sexual Cycle in the Human Female as Revealed by Vaginal Smears." *American Journal of Anatomy* 52, no. May (1933): 519–637.

Papanicolaou, George, and Herbert Traut. *Diagnosis of Uterine Cancer by the Vaginal Smear*. New York: Commonwealth Fund, 1943.

Papanicolaou, George, and Herbert Traut. "The Diagnostic Value of Vaginal Smears in Carcinoma of the Uterus." *American Journal of Obstetrics and Gynecology* 42, no. 2 (1941): 193–206.

"Papanicolaou Testing—Are We Screening the Wrong Women?" Editorial. *New England Journal of Medicine* 294 (1976): 223.

Park, Katherine. "Nature in Person: Renaissance Allegories and Emblems." In *The Moral Authority of Nature*. Edited by Lorraine Daston and Fernando Vidal, 50–73. Chicago: University of Chicago Press, 2003.

Pascoe, Peggy. *Relations of Rescue: The Search for Female Moral Authority in the American West, 1874–1939*. Oxford: Oxford University Press, 1990.

Patterson, James. *The Dread Disease: Cancer and Modern American Culture*. Cambridge: Harvard University Press, 1987.

Pauly, Philip. *Controlling Life: Jacques Loeb and the Engineering Ideal in Biology*. Oxford: Oxford University Press, 1987.

Payne, Charles. *I've Got the Light of Freedom: The Organizing Tradition and the Mississippi Freedom Struggle*. Berkeley: University of California Press, 1995.

Pearl, Raymond, and Agnes Bacon. "The Racial and Age Incidence of Cancer and of Other Malignant Tumors." *Archives of Pathology and Laboratory Medicine* 3 (1927): 963–92.

Pereyra, A. J. "The Relationship of Sexual Activity to Cervical Cancer: Cancer of the Cervix in a Prison Population." *Obstetrics and Gynecology* 17, (February 1961): 154–59.

Perlstein, Daniel. "Teaching Freedom: SNCC and the Creation of the Mississippi Freedom Schools." *History of Education Quarterly* 30, no. 3 (1990): 297–324.

Petchesky, Rosalind. *Global Prescriptions: Gender, Health and Human Rights*. London: Zed Press, 2003.

Petryna, Adriana. *Life Exposed: Biological Citizens after Chernobyl*. Princeton: Princeton University Press, 2002.

Petryna, Adriana, Andrew Lakoff, and Arthur Kleinman, eds. *Global Pharmaceuticals: Ethics, Markets, Practices*. Durham: Duke University Press, 2006.

Pigg, Stacy Leigh. "Globalizing the Facts of Life." In *Sex in Development: Science, Sexuality and Morality in Global Perspective*. Edited by Stacy Leigh Pigg and Vincanne Adams, 39–65. Durham: Duke University Press, 2005.

Pityana, Barney, ed. *Bounds of Possibility: The Legacy of Steve Biko and Black Consciousness*. London: Zed, 1991.

Polletta, Francesca. *Freedom Is an Endless Meeting*. Chicago: Chicago University Press, 2002.

The Proceedings of the Menstrual Extraction Conference. Oakland: Oakland Feminist Women's Health Center, 1974.

Pulido, Laura. *Black, Brown, Yellow and Left: Radical Activism in Los Angeles*. Berkeley: University of California Press, 2006.

Rabinow, Paul. "Artificiality and Enlightenment: From Sociobiology to Biosociality." In *Incorporations*. Edited by Jonathan Crary and Sanford Kwinter, 99–103. New York: Zone, 1992.

Rajan, Kaushik Sunder. *Biocapital: The Constitution of Postgenomic Life*. Durham: Duke University Press, 2006.

Rajan, Tilottama. "Dis-Figuring Reproduction: Natural History, Community, and the 1790s Novel." *The New Centennial Review* 2, no. 3 (2002): 211–52.

Ramazanoglu, Carolyn, ed. *Up against Foucault: Explorations of Some Tensions between Foucault and Feminism*. London: Routledge, 1993.

Ramsden, Edmund. "Carving up Population Science: Eugenics, Demography and the Controversy over the 'Biological Law' of Population Growth." *Social Studies of Science* 32, nos. 5–6 (2002): 857–99.

———. "Social Demography and Eugenics in the Interwar United States." *Population and Development Review* 29, no. 4 (2003): 547–93.

Randall, C. L. "The Obstetrician-Gynecologist and Reproductive Health." *American Journal of Obstetrics and Gynecology* 129, no. 7 (1977): 715–22.

Ransby, Barbara. *Ella Baker and the Black Freedom Movement: A Radical Democratic Vision*. Chapel Hill: University of North Carolina Press, 2003.

Rapp, Rayna, Deborah Heath, and Karen-Sue Taussig. "Flexible Eugenics: Technologies of the Self in the Age of Genetics." In *Genetic Nature/Culture: Anthropology and Science Beyond the Two-Culture Divide*. Edited by Alan Goodman, Deborah Heath, and M. Susan Lindee, 51–76. Berkeley: University of California Press, 2003.

Ravenholt, Reimert. "Aid's Family Planning Strategy." *Science* 163, no. 3863 (1969): 124–27.

———. "The Power of Availability." Paper presented at the Village and Household Availability of Contraceptives Conference, Tunis, Tunisia, 1977.

———. "Taking Contraceptives to the World's Poor: Creation of USAID's Population / Family Planning Program, 1965–80," unpublished manuscript, www.ravenholt.com/contraceptives.

Ravenholt, Reimert, and D. G. Gillespie. "Maximizing Availability of Contraception through Household Utilization." In *Village and Household Availability of Contraceptives: Southeast Asia*. Edited by J. Gardner, R. Wolff, D. Gillespie, and G. Duncan. Seattle: Battelle Human Affairs Research Centers, 1976.

Reagan, Leslie. "Crossing the Border for Abortions: California Activists, Mexican Clinics, and the Creation of a Feminist Health Agency in the 1960s." *Feminist Studies* 26, no. 2 (2000): 323–48.

————. *When Abortion Was a Crime: Women, Medicine, and Law in the United States, 1867–1973*. Berkeley: University of California Press, 1997.

Redstockings, ed. *Feminist Revolution*. New York: Random House, 1975.

Reeves, W. C., L. A. Brinton, and M. Garcia. "Human Papillomavirus Infection and Cervical Cancer in Latin America." *New England Journal of Medicine* 320 (1989): 1437–41.

Report of the President's Cancer Panel. The Meaning of Race in Science: Considerations for Cancer Research. Bethesda: National Institutes of Health. National Cancer Institute, 1998.

Reverby, Susan. "Alive and Well in Somerville, Mass." *HealthRight* 1, no. 2 (1975): 1.

Rheinberger, Hans-Jörg. *Toward a History of Epistemic Things: Synthesizing Proteins in the Test Tube*. Stanford: Stanford University Press, 1997.

Ries, L. A., et al., eds. *Seer Cancer Statistics Review 1973–1995*. Bethesda: National Cancer Institute, 1998.

Rigdon, R. H., Helen Kirchhoff, and Mary Lee Walker. "Frequency of Cancer in the White and Colored Races as Observed at the Autopsy between 1920 and 1949 at the Medical Branch." *Texas Reports on Biology and Medicine* 10 (1952): 914–28.

Roberts, Dorothy. *Killing the Black Body: Race, Reproduction, and the Meaning of Liberty*. New York: Vintage, 1997.

Rodriguez-Trias, Helen. "The Women's Health Movement: Women Take Power." In *Reforming Medicine: Lessons of the Last Quarter Century*. Edited by Victor Sidel, 107–26. New York: Pantheon Books, 1984.

Rofel, Lisa. *Desiring China: Experiments in Neoliberalism, Sexuality, and Public Culture*. Durham: Duke University Press, 2007.

Rogers, Carl. *On Encounter Groups*. New York: Harper and Row, 1970.

————. *A Plan for Self-Directed Change in an Educational Institute*. La Jolla, Calif.: Western Behavioral Science Institute, 1967a.

————. "The Process of the Basic Encounter Group." In *Challenges of Humanistic Psychology*. Edited by J. F. T. Bugental, 261–76. New York: McGraw-Hill, 1967b.

Rogers, Naomi. "'Caution: The AMA May Be Dangerous to Your Health': The Student Health Organizations (SHO) and American Medicine, 1965–1970." *Radical History Review* 80 (2001): 5–34.

Rolinson, Mary. *Grassroots Garveyism: The Universal Negro Improvement Association in the Rural South*. Chapel Hill: University of North Carolina Press, 2007.

Rose, Hilary. "Hand, Brain, and Heart: A Feminist Epistemology for the Natural Sciences." *Signs* 9, no. 1 (1983): 73–90.

Rose, Nikolas. "Governing 'Advanced' Liberal Democracies." In *Foucault and Political Reason: Liberalism, Neo-Liberalism and Rationalities of Government*. Edited by Andrew Barry, Thomas Osborne, and Nikolas Rose, 37–64. Chicago: University of Chicago Press, 1996.

————. *The Politics of Life Itself: Biomedicine, Power, and Subjectivity in the Twenty-First Century*. Princeton: Princeton University Press, 2007.

Ross, C. E., J. Mirowsky, and W. C. Cockerham. "Social Class, Mexican Culture, and Fatalism: Their Effects on Psychological Distress." *American Journal of Community Psychology* 11, no. 4 (1983): 388–99.

Ross, Walter. *Crusade: The Official History of the American Cancer Society*. New York: Arber House, 1987.

Rossinow, Doug. *The Politics of Authenticity: Liberalism, Christianity and the New Left in America*. New York: Columbia University Press, 1998.

Rothman, Lorraine. *The Proceedings of the Menstrual Extraction Conference*. San Francisco/Oakland: Oakland Feminist Women's Health Center, 1974.

Rudd, Peter. "The United Farm Workers Clinic in Delano, Calif.: A Study of the Rural Poor." *Rural Health* 90, no. 4 (1975): 331–39.

Russell, Barbara, and Lynn Lofstrom. "Health Clinic for the Alienated." *American Journal of Nursing* 71, no. 1 (1971): 80–83.

Rutherford, Charlotte. "Reproductive Freedoms and African American Women." *Yale Journal of Law and Feminism* 4, no. 2 (1992): 255–190.

Saikia, Yasmin. *Women, War and the Making of Bangladesh: Remembering 1971*. Durham: Duke University Press, 2011.

Sandoval, Chela. *Methodology of the Oppressed*. Minneapolis: University of Minnesota Press, 2000.

———. "New Sciences: Cyborg Feminism and the Methodology of the Oppressed." In *The Cyborg Handbook*. Edited by C. H. Gray. New York: Routledge, 1995.

Sanger, Margaret. "The Birth Control League." *The Woman Rebel* 1, no. 5 (1914): 7.

Sankaranarayanan, R., et al. "Visual Inspection of the Uterine Cervix after the Application of Acetic Acid in the Detection of Cervical Carcinoma and Its Precursors." *Cancer* 83, no. 10 (1998): 2150–56.

Sarachild, Kathie. "Consciousness-Raising: A Radical Weapon," in *Redstockings, Feminist Revolution*, 144–50. New York: Random House, 1975.

Schabas, Margaret. "Adam Smith's Debts to Nature." *History of Political Economy* 35 (2003): 262–81.

———. *The Natural Origins of Economics*. Chicago: University of Chicago Press, 2005.

Schaffer, Simon. "Self Evidence." *Critical Inquiry* 18, no. 2 (1992): 327–62.

Schiebinger, Londa. "Why Mammals Are Called Mammals: Gender Politics in Eighteenth Century Natural History." *American Historical Review* 98 (1993): 382–411.

Schlosser, Kolson. "Bio-Political Geographies." *Geography Compass* 2, no. 5 (2008): 1621–34.

Schmidt, O. "Cervical Cancer Screening Program: The SOGC's View." *Canadian Medical Association Journal* 116 (1977): 971–75.

Schoen, Johanna. *Choice and Coercion: Birth Control, Sterilization, and Abortion in Public Health and Welfare*. Chapel Hill: University of North Carolina Press, 2005.

Schultz, Susanne. "Dissolved Boundaries and 'Affective Labor': On the Disappearance of Reproductive Labor and Feminist Critique in *Empire*." *Capitalism, Nature, Socialism* 17, no. 1 (2006): 77–82.

Scott, David. "Colonial Governmentality." *Social Text* 43, no. (autumn 1995): 191–220.

———. "Counting Women's Caring Work: An Interview with Andaiye." *Small Axe* 8, no. 1 (2004): 123–217.

Scott, Joan. *Only Paradoxes to Offer: French Feminists and the Rights of Man*. Cambridge: Harvard University Press, 1996.

———. "The Evidence of Experience." *Critical Inquiry* 17, (summer 1991): 773–97.

Shah, Nayan. *Contagious Divides: Epidemics and Race in San Francisco's Chinatown*. Berkeley: University of California Press, 2001.

Shakry, Omnia El. *The Great Social Laboratory: Subject of Knowledge in Colonial and Post-colonial Egypt*. Stanford: Stanford University Press, 2007.

Shapin, Steven, and Simon Schaffer. *Leviathan and the Air-Pump: Hobbes, Boyle, and the Experimental Life*. Princeton: Princeton University Press, 1985.

Shim, Janet. "Constructing 'Race' across the Science-Lay Divide: Racial Formation in the Epidemiology and Experience of Cardiovascular Disease." *Social Studies of Science* 35, no. 3 (2002): 405–36.

Siddall, A. Clair. "The Gynecologist's Role in Comprehensive Medical Care." *American Journal of Public Health* 59, no. 4 (1969): 657–62.

———. "The Presumably Well Woman." *American Journal of Obstetrics and Gynecology* 64, no. 1 (1952): 168–73.

———. "Preventative Medicine in Gynecologic Practice." *Obstetrics and Gynecology* 12, no. 2 (1958): 230–32.

Silliman, Jael. "Expanding Civil Society: Shrinking Political Spaces. The Case of Women's NGOs." In *Dangerous Intersections: Feminist Perspectives on Population, Development and the Environment*. Edited by Ynestra King and Jael Silliman, 23–53. Cambridge, Mass.: South End Press, 1998.

Silliman, Jael, et al. *Undivided Rights: Women of Color Organize for Reproductive Justice*. Cambridge, Mass.: South End Press, 2004.

Silver, Morton. "Birth Control and the Private Physician." *Family Planning Perspectives* 4, no. 2 (1972): 42–26.

Simonds, Wendy. *Abortion at Work: Ideology and Practice in a Feminist Clinic*. New Brunswick: Rutgers University Press, 1996.

Sister Song Women of Color Reproductive Health Collective. *Reproductive Justice Briefing Book*. Atlanta: Sister Song, 2007.

Slawson, D. C., J. H. Bennett, and J. M. Herman. "Are Papanicolaou Smears Enough? Acetic Acid Washes of the Cervix as Adjunctive Therapy: A Harnet Study." *Journal of Family Practice* 35, no. 3 (1992): 271–77.

Smiles, Samuel. *Self Help*. London: John Murray, 1882.

Smith, Andrea. *Conquest: Sexual Violence and American Indian Genocide*. Cambridge, Mass.: South End Press, 2005.

Smith, Barbara. "'Feisty Characters' and 'Other People's Causes': Memories of White Racism and U.S. Feminism." In *The Feminist Memoir Project: Voices from Women's*

Liberation, edited by Rachel Blaue Du Plessis and Ann Snitow, 477–81. New York: Crown Publishing, 1998.

Smith, Beverly. "Black Women's Health: Notes for a Course." In *But Some of Us Are Brave: Black Women's Studies*. Edited by Gloria T. Hull, Patricia Bell Scott, and Barbara Smith, 103–14. Boston: Feminist Press, 1982.

———. "The Development of an Ongoing Data System at a Women's Health Center." MA thesis. University of Massachusetts, Boston, 1976.

Smith, Dorothy. "Some Implications of a Sociology for Women." In *Woman in a Man-Made World: A Socioeconomic Handbook*. Edited by Nona Glazer and Helen Waehrer, 15–29. Chicago: Rand-McNally, 1977.

———. "Women's Perspective as a Radical Critique of Sociology." *Sociological Inquiry* 44 (1974): 7–14.

Smith, Susan. *Sick and Tired of Being Sick and Tired: Black Women's Health Activism in America*. Philadelphia: University of Pennsylvania Press, 1995.

Smith-Rosenberg, Carroll. *Disorderly Conduct: Visions of Gender in Victorian American*. New York: A. A. Knopf, 1985.

Soderstrom, Richard. "Menstrual Regulation Technology." In *Pregnancy Termination: PARFR Series on Fertility Regulation*. Edited by G. I. Zatuchni, J. J. Sciarra, and J. J. Speidel, 60–68. Hagerstown, Md.: Harper and Row, 1979.

Solinger, Rickie. *Beggars and Choosers: How the Politics of Choice Shapes Adoption, Abortion, and Welfare in the United States*. New York: Hill and Wang, 2002.

———. *Wake up Little Susie: Single Pregnancy and Race before Roe v. Wade*. New York: Routledge, 2000.

Speich, Daniel. "The World of GDP: Historicizing the Epistemic Space of Postcolonial Development." Paper presented at the "Borders and Boundaries—Grenzüberschreitungen—Geschichte—Globale Gleichzeitigkeit," Centro Stefano Franscini, Ascona, June 3–8, 2007.

Spivak, Gayatri Chakravorty. "In a Word: Interview." In *Outside in the Teaching Machine*, 1–24. New York: Routledge, 1993.

Sretzer, Simon. "The Idea of Demographic Transition and the Study of Fertility: A Critical Intellectual History." *Population and Development Review* 19, no. 4 (1993): 659–701.

Stanczuk, Grazyna A., et al. "Typing of Human Papillomavirus in Zimbabwean Patients with Invasive Cancer of the Uterine Cervix." *Acta Obstetricia et Gynecologica Scandinavica* 82, no. 8 (2003): 762–66.

Stern, Alexandra Minna. "Buildings, Boundaries, and Blood: Medicalization and Nation-Building on the U.S.-Mexico Border, 1910–1930." *Hispanic American Historical Review* 79, no. 1 (1999): 41–81.

———. *Eugenic Nation: Faults and Frontiers of Better Breeding in Modern America*. Berkeley: University of California Press, 2005a.

———. "Sterilized in the Name of Public Health: Race, Immigration, and Reproductive Control in Modern California." *American Journal of Public Health* 95, no. 7 (2005b): 1128–38.

Stoler, Ann Laura. *Carnal Knowledge and Imperial Power: Race and the Intimate in Colonial Rule*. Berkeley: University of California Press, 2002.

———. *Race and the Education of Desire: Foucault's History of Sexuality and the Colonial Order of Things*. Durham: Duke University Press, 1995.

———. "Racial Histories and Their Regimes of Truth." *Political Power and Social Theory* 11 (1997): 183–206.

Summers, Lawrence. "The Most Influential Investment," *Scientific American* 267, no. 2 (1992): 108,132.

Sung, Hung-chao. "A Revolutionized Mind Is the Key to Success in Research Work." *Chinese Journal of Obstetrics and Gynecology* 11, no. 6 (1965): 389.

Sung, John, Ernest Alema-Mensah, and Daniel Blumenthal. "Inner-City African American Women Who Failed to Receive Cancer Screening Following a Culturally-Appropriate Intervention: The Role of Health Insurance." *Cancer Detection and Prevention* 26 (2002): 28–32.

Sung, John, et al. "Cancer Screening Intervention among Black Women in Inner-City Atlanta: Design of a Study." *Public Health Reports* 107, no. 4 (1992): 381–88.

Taguieff, Pierre-Andre. "The New Cultural Racism in France." *Telos* 83 (1990): 109–22.

Terranova, Tiziana. "Free Labor: Producing Culture for the Digital Economy." *Social Text* 18, no. 2 (2000): 33–59.

Thayer, Millie. "Transnational Feminism: Reading Joan Scott in the Brazilian Sertao." *Ethnography* 2 (2001): 243–71.

Thomas, Jan, and Mary Zimmerman. "Feminism and Profit in American Hospitals: The Corporate Construction of Women's Health Centers." *Gender and Society* 21, no. 3 (2007): 359–83.

Thompson, Charis. *Making Parents: The Ontological Choreography of Reproductive Technologies*. Cambridge: MIT Press, 2005.

Tone, Andrea. *Devices and Desires: A History of Contraception in America*. New York: Hill Wang, 2001.

Torroella-Kouri, M., et al. "HPV Prevalence among Mexican Women with Neoplastic and Normal Cervixes." *Gynecological Oncology* 70, no. 1 (1998): 115–120.

Traweek, Sharon. *Beamtimes and Lifetimes*. Cambridge: Harvard University Press, 1988.

Trotz, Alissa. "Red Thread: The Politics of Hope in Guyana." *Race and Class* 49, no. 2 (2007): 71–79.

Trumbull, Robert. "Dacca Raising the Status of Women While Aiding Rape Victims." *New York Times*, May 12, 1972, 2.

Tuana, Nancy. "Coming to Understand: Orgasm and the Epistemology of Ignorance." *Hypatia: A Journal of Feminist Philosophy* 19, no. 1 (2004): 194–232.

UBINIG. *Violence of Population Control*. Dhaka: Narigrantha Prabartana, 1999.

University of Zimbabwe / JHPIEGO Cervical Cancer Project. "Visual Inspection with Acetic Acid for Cervical-Cancer Screening: Test Qualities in a Primary-Care Setting." *Lancet* 353, no. 9156 (1999): 869–73.

Vaillant, H. W., G. T. Cummins, and R. M. Richart. "An Island-Wide Screening Program for Cervical Neopolasia in Barbados." *American Journal of Obstetrics and Gynecology* 101, no. 7 (1968): 943–46.

Vancouver Women's Health Collective. *A Feminist Approach to Pap Tests*. Vancouver: Press Gang, 1986.

Vettel, Eric. *Biotech: The Countercultural Origins of an Industry*. Philadelphia: University of Pennsylvania Press, 2006.

Villa, L. L., and E. L. Franco. "Epidemiologic Correlates of Cervical Neoplasia and Risk of Human Papillomavirus Infection in Asymptomatic Women in Brazil." *Journal of the National Cancer Institute* 81 (1991): 332–40.

Wailoo, Keith. *Dying in the City of Blues: Sickle Cell Anemia and the Politics of Race and Health*. Chapel Hill: University of North Carolina Press, 2001.

———. *When Cancer Crossed the Color Line*. Oxford: Oxford University Press, 2011.

Wailoo, Keith, Julie Livingston, Steven Epstein, and Robert Aronowitz, eds. *Three Shots at Prevention: The HPV Vaccine and the Politics of Medicine's Simple Solutions*. Baltimore: Johns Hopkins University Press, 2010.

Walboomers, Jan, et al. "Human Papillomavirus Is a Necessary Cause of Invasive Cervical Cancer Worldwide." *Journal of Pathology* 189, no. 1 (1999): 12–19.

Waldby, Catherine. "Stem Cells, Tissue Cultures and the Production of Biovalue." *Health* 6, no. 3 (2002): 305–23.

Walton, J. R., et al. "Cervical Cancer Screening Programs." *Canadian Medical Association Journal* 114 (1976): 1003–33.

Ward, Martha. *Poor Women, Powerful Men: American's Great Experiment in Family Planning*. Boulder: Westview Press, 1986.

Ward, Stephen. "The Third World Women's Alliance: Black Feminist Radicalism and Black Power Politics." In *The Black Power Movement: Rethinking the Civil Rights–Black Power Era*. Edited by Peniel Joseph, 119–44. New York: Routledge, 2006.

Wasserheit, Judith. "The Significance and Scope of Reproductive Tract Infections among Third World Women." *International Journal of Gynecology and Obstetrics* 30, no. suppl. 3 (1989): 145–68.

Watkins, Elizabeth. *On the Pill: A Social History of Oral Contraceptives, 1950–1970*. Baltimore: Johns Hopkins University Press, 1998.

"Watts Riots Anniversary: Community Fights to Save Jobs, Watts Health Foundation Created in Response to Violence." *Los Angeles Sentinel*, August 14, 2002.

West Coast Sisters. "How to Start Your Self-Help Clinic, Level II." Mimeograph. Los Angeles, 1971.

White, Deborah Gray. *Too Heavy a Load: Black Women in Defense of Themselves, 1894–1994*. New York: W. W. Norton and Company, 1999.

White, Sarah. "Thinking Race, Thinking Development." *Third World Quarterly* 23, no. 3 (2002): 407–19.

White, Tyrene. *China's Longest Campaign: Birth Planning in the People's Republic, 1949–2005.* Ithaca: Cornell University Press, 2006.

Whyte, William. *The Organizational Man.* Garden City: Doubleday Anchor Books, 1956.

Williams, D. "Race/Ethnicity and Socioeconomic Status: Measurement and Methodological Issues." *International Journal of Health Services* 26 (1996): 483–505.

Williams, D., R. Lavizzo-Mourey, and R. C. Warren. "The Concept of Race and Health Status in American." *Public Health Reports* 109 (1994): 26–41.

Williamson, A. L., et al. "Typing of Human Papillomaviruses in Cervical Carcinoma Biopsies from Cape Town." *Journal of Medical Virology* 43, no. 3 (1994): 231–37.

Winant, Howard. "Behind Blue Eyes: Whiteness and Contemporary U.S. Racial Politics." *New Left Review* 225 (September/October 1997): 73–89.

Women and Development Unit. "Demystifying and Fighting Cervical Cancer." *Isis International Women's Health Journal* 92, no. 3 (1992): 29–51.

Women's Community Health Center. *Annual Report.* Cambridge, Mass., 1978.

"Women's Voices '94: A Declaration on Population Policies." *Population and Development Review* 19, no. 3 (1993): 637–40.

Worcester, Nancy, and Mariamne Whatley. "The Response of the Health Care System to the Women's Health Movement: The Selling of Women's Health Centers." In *Women's Health: Readings on Social, Economic, and Political Issues.* Edited by Nancy Worcester and Mariamne Whatley, 17–23. Dubuque: Kendall/Hunt, 1988a.

———. "The Role of Technology in the Co-optation of the Women's Health Movement: The Case of Breast Cancer Screening." In *Women's Health: Readings on Social, Economic, and Political Issues.* Edited by Nancy Worcester and Mariamne Whatley, 187–92. Dubuque: Kendall, 1988.

World Health Organization. *World Cancer Report.* Geneva: International Agency for Research on Cancer, 2003.

World Health Organization. *World Cancer Report.* Geneva: International Agency for Research on Cancer, 2005.

Wright, Melissa. *Disposable Women and Other Myths of Global Capitalism.* New York: Routledge, 2006.

Wu, Y. T., and H. C. Wu. "Suction in Artificial Abortion: Report of 300 Cases." *Chinese Journal of Obstetrics and Gynecology* 6, no. 1 (1958): 447–49.

Wu, Yuan-t'ai, and Ling-mei Shang. "Clinical Application of the Negative Pressure Bottle." *Chinese Journal of Obstetrics and Gynecology* 11, no. 6 (1965): 390–93.

Ziff, R. "The Role of Psychographics in the Development of Advertising Strategy and Copy." In *Life Style and Psychographics.* Edited by D. W. Wells. Chicago: American Marketing Association, 1974.

Page numbers in italics refer to illustrations.

abortion: in China, 155, *156*, 157; historically, 214n10; illegal, 46, 155–57, 168,
169, 215n10; legislating, 18, 168, 169,
215n10, 215n13; Menstrual Extraction
(ME) vs., 158–59; protocols, 28, 190n6;
for raped women, 168; responsibiliza-
tion of women and, 183n1; safety, 157;
self help group formation and, 190n7;
self-performed, 46, 157; statistics,
215n10
activism, 20, 41
affect, category of, 92–93
affective economies: of capital and desire,
91–93; politics of immodest witnessing
and, 79–80; of technoscience, 72–73; of
vaginal self-examination, 80–90
affective labor, 92
Agamben, Giorgio, 44
agribusiness, 1, 5
Ahmed, Sara, 79–80
AIDS, 130
Aid to Needy Children (ANC), 44
Akhter, Farida, 7
Allen, Pamela, 84
Alma Alta Declaration (WHO), 51
American Association of Cytopathology,
147
American Cancer Society (ACS), 110, 112,
123

American College of Obstetrics and Gyne-
cologists, 110
American Society for the Control of Can-
cer, 107, 110
Andaiye, 140–41, 145
animal reproduction, 1
Annan, Kofi, 144
antiabortion movement, 64, 66, 100
Antrobus, Peggy, 139
Army of Three, 190n7
Asian feminists, 55
Asian Pacific health activists, 55
Asian Women's Health Project, 55
Avery, Byllye, 44, 65

Bangladesh, 164–65, 168, 169, 172
Bangladesh Women's Health Coalition
(BWHC), 174
Barbados, 141
Barbados Family Planning Association,
148, 207n52
Beasley, Joseph, 188n40
Beaton, Brian, 8, 46
Beauvoir, Simone de, 9
Berg, Marc, 192n11
Bergeron, Suzanne, 17
Biko, Steve, 58
biocapital, 19–20
biocitizenship, 20–21

biomedicalization: constraints imposed by, 48; emergence, features of, 19–20; periodization, 196n73; politics of health and, 50–56

biomedical logics, 52

biomedicine: creation of, 71; emergence of, 19, 71; entanglements, 102–3, 116, 129; feminist reconfiguring of, 33, 48, 72; industrialization of, 19–20; moral economy of, 71; points of interaction for women, 113–14, 131; as a political economy, 133–34, 149; politicization of, 1–2; protocols, 29; race and, 124, 128, 130–33

biopolitical project of feminism, 9–10

biopolitical topology, 11–20, 44–45, 187n25

biopolitical we, 37–38

biopolitics: entanglements, 36–37; feminism as, 24; of feminist self help, 36–40, 65; Foucault's formulation of, 11, 13, 24, 36–37, 175; of Menstrual Extraction (ME), 161, 163; necropolitical dimension, 43–44; twentieth century, 175

biosocialities, 57

biotechnology, 72

biovalue, 97–98

birth control, 8; movement for, 33; the pill as, 16,20

birth planning, China, 155

Black Book, 65, 191n8

black consciousness, 58

black feminists, 34, 55, 65–66

"A Black Feminist Statement" (Combahee River Collective), 38–39, 43

black genocide, 44

Black Panthers, 5, 55

Black Power movement, 58

black radicalism, transatlantic, 58

blacks: and disease, racialization of, 107, 124–25; health care for, 128, 131–33; health status improvements, 131;

women's health projects, 65–66, 125–26, 128–33

black women physicians, 129

Black Women's National Health Project (BWNHP), 65

Blue Lady Oral Contraceptives, 164

bodies, politicization of, 9–10

Boston, feminist self-help in, 38–40, 42–44

Boston Women's Health Book Collective, 38, 89

Brown, Laura, 171–72, 191n9

Buffon, Georges-Louis Leclerc, comte de, 185n17

Bunchy Carter Free Medical Clinic, 55

Bush, George W., 146

Butler, Judith, 87

Calcutta Telegraph, 147

California Board of Medical Quality Assurance, 63–64

Cambrosio, Albert, 71

Canada: cervical cancer rates, 134; feminist health centers, 112–13, 128; feminist self help in, 27; Pap smear screening programs, 102, 104, 105, 106, 112, 120, 122–23, 126, 131, 135, 207n52, 208n57; radical feminist health collectives, 115; universal health care, 122, 123

Canadian Society of Gynecologists and Obstetricians, 123

Canadian Women's Health Network, 148

cancer: racializations and genderings of, 106–7; redefining, pathologists in, 107–8. See also cervical cancer; Pap smear screening

cancer cells classification system, 111, 112

Cancer Journals (Lorde), 133, 141

capitalism: affective entanglements with, 91–93; biomedicalization of, 19–20; entanglements, 7, 15; reassembly of, 19–20; reproduction-production relations, 7, 8, 11; sexed-labor division in, 5

Carolina Population Center, 169

Casper, Monica, 102, 110

cervical cancer: causes of, 134, 136; commodification of risk, 147; feminist mapped interventions, 105, 135–49; historically, 106–7; mortality, 106, 135, 146; ontological politics of, 103–4, 115, 133–35, 136, 149; politicization, 136, 138–43; risk of, 136, 138–41, 147; transnational scale of, 135; treatment of, 106

cervical cancer, problem-space of: economic, 105, 121–23, 126, 128, 131–32; feminist remapping, 105, 135–49; political economy, 125–27, 133–34; racial, 105–7, 123–28, 131–32; solution of the Pap smear in, 104

cervical cancer research, sites of, 134

cervix, 96, 103, 119

change agents, 57, 59, 60, 61

Chatterjee, Partha, 20

Chernobyl, 20

Chicana feminists, 34, 55, 163

Chicana (film), 96

China, suction abortion in, 155, 157

Chinese Revolution, 58

Christian liberation theology, 58

citizenship: biological, 20–21; fit, defining, 130; racialized forms of, 44, 131; universal, 9

civil rights, 50–51; legislation for, 121–22

Clark, Sawin, 110

Clarke, Adele, 19, 52, 102, 108

Clinical Research Organization (CRO), 170–71

clinical trials, 20, 170–71

Clinton, Bill, 146

Clinton, Hillary, 143, 146

clitoris, anatomy and function of, 78–79

Cohen, Lawrence, 35

Cold War: biopolitics and reproduction in, 15–16, 153, 154, 163; human relations methods in, 59

Collins, Patricia Hill, 98

colonialism, justification for, 16

Combahee River Collective, 38–39, 40, 42–44, 45–46, 84–85

Comilla, Bangladesh, 7

Committee to End Sterilization Abuse, 45

communism, birth control and, 15

community health centers, 55

Computer Lib (Nelson), 39

conception, time of, 16

consciousness raising, 80–85; architects of, 201n24; "click" sound of, 47–48; decolonizing the mind, 58; experience's role in; power of the vaginal self-exam in, 47, 100–101; protocols of, 26, 88; small group work and, 57; social change through, 62

contraception distribution: decriminalization of, 18; USAID strategy for, 20, 164–68, 166

counter-conduct: affective entanglements form of, 90–93; feminist forms of, 29; feminist self help as a project of, 37, 49, 61–62, 64; health as a domain of, 54–56; medical profession, 55; resistance vs., 183n3; technoscientific, 4, 72; term usage, 184n3, 192n10; vaginal self-examination as a form of, 49, 72, 99–100

counter topographies, 100, 186n23

Cruikshank, Barbara, 62, 138

cultural racism, 124

cytology, clinical field of, 110, 112

cytopathology profession, 147

cytotechnicians, 121

Daston, Lorraine, 69, 70, 72, 79, 80

Davidson, Arnold, 183n3, 192n10

Davis, Angela, 132, 133

De Beers Group, 144

Decade for Women (UN), 17

Declaration of Comilla, 7

decolonization, psychic, 59

Del-Em Menstrual Extraction device, 159

Delueze, Gilles, 29

democracy: birth control in sustaining, 15; defined, 59; small group practices and, 59–60

demographics, demography, 17, 91

desire: in fertility and consumption, 165, 167–68; marketing strategies of, 91–93

"The Development of an Ongoing Data System at a Women's Health Center" (Smith), 39

The Dialectic of Sex (Firestone), 5

dilation and curettage, 214n10

double consciousness, 23

Downer, Carol, 46, 48, 49, 50, 60–63, 81, 82, 157, 160, 172, 190nn6, 7, 199n125

Draper, William, 15

Dreifus, Claudia, 34

Du Bois, W. E. B., 23

Dunlop, Joan, 135–36

Duster, Troy, 124

Echols, Alice, 191n9

ecology of vagina, 94–95, 97, 119

economic development: centrality of women in, 17; incorporation of fertility into, 3–4, 17–19, 53, 153–54, 164; reproductive health and, 143–46; service and knowledge sectors, 62

economics: of affect, 91–93; field of, 17; of health, 20, 51–52, 121–23

economization: of fertility, 3–4, 17–19, 53, 153–54, 164; of life as human capital, 51, 62

economy of nature, 9

education of women, 17–18

Ehrenreich, Barbara, 171

Eisenhower, Dwight D., 15

emotions, 79–80; emotional labor and, 92

empowerment: injunctions to, 61; meaning of, 23; philosophy of, in health care, 128; politics of, 143–45; as term, 138; will to, 62; of women in reproductive health, 135, 138–44

encounter groups, 58–59, 61–62

Engels, Friedrich, 5

entanglements: of biomedicine, 102–3, 116, 129; biopolitical, 36–37, 154, 170–75; of capitalism, 7, 15; defined, 12; of economies, 53, 93; examples of, 13; feminisms, 2–3, 6; of feminist self help, 30, 32–33, 62, 66, 93, 116; Menstrual Extraction and, 152–53, 171–75; Menstrual Regulation and, 152–53; the Pap smear, 102, 112, 120, 127, 148; of technoscience, 2, 4–5, 7, 36; women's health movement, 19

entanglements, affective: of capitalism and desire, 91–93; of vaginal self-examination, 57, 72–73, 79, 89–90

environmental science, 5

eugenics, 9–11, 18, 130; flexible, 183n1 legislation for, 53

existentialism, 58

experience, 80–86, 98–99

Family Health International, 170

family planning: Cold War and militarization of, 15–17, 163–70; economic development and, 53; foreign aid funding for, 16–17, 163–70, 166, 174; necropolitical effect, 16; neoliberal, elements shaping, 175; racial and economic stratification, 53; the state and, 16, 18–19, 53–54

family planning: clinics for, 4; infrastructure for, 169–71; NGOs for, 20, 169

Fanon, Franz, 16, 55, 58

Federation of Feminist Health Centers, 40

Federation of Feminist Women's Health Centers (FFWHC), 86

Federation of Women's Health Centers, 64, 190n8

feminism: biopolitical project of, 9–10; biopolitical topography, 14–21; as biopolitics, meaning of, 24; emergence of, 14–15; NGO-ization of, 6, 31, 66, 136–37, 146, 174; tools necessary to, 5 feminisms, feminists: categorization of, 29; concerns primary to, 10; consciousness raising and, 62; contradictions informing, 19, 45; counter-conduct and, 29; defining features, 2–3; entanglements, 2–3, 6; formations of, 29; formations of racialization, 41–43; politics of, 2, 14; practices of care of, 176; projects of, 7; revolution of, 5–6; shaping of, 40; technoscience and, 31; "woman" meaning in, 14

A Feminist Approach to Pap Tests (booklet), 125–26

Feminist Economic Network (FEN), 191n9

feminist health care: centers and clinics of, 2, 15, 63–64, 113–14, 116, 128–29; NGOS and, 138, 174; space of, 116

feminist health movement, 129; projects of, 3, 8, 10

feminist health politics: biopolitical topology, 15–16; Boston, 38–40, 42–44; elements shaping, 114; simultaneous yet discrepant, 38–44

feminist self help: advanced research groups, 27, 79, 86, 93; affective economy of, 80–81; antiauthoritarian values in, 116, 118–19; anticolonial rhetoric, 34; beginnings, 190n7; biopolitics, 36–40, 49, 65, 68, 72, 85–90; Boston version, 38–40; contradictions informing, 34; counter-empire of, 171–72; double vision of, 32; emergence of, 39–41, 53; entanglements, 30, 116; focus of, 33, 39, 42–46; as a form of protocol feminism, 28–31; founding narrative of, 46–49; groups for, 27, 82–84, 88–89; human relations techniques, 62; iconic artifacts, 30; ideology, 33–34, 35, 115; information-sharing in, 116, 118; kit for, 25; labor politics, 62–63; limits of, 115–16; Los Angeles version, 40; moral economy of affirmation, 72–73, 79, 89–90; NGO-ization of, 66; operability ethic, 35; politicization of, historically,

35–36; practices, 30; protocol politics, 115–16, 118; protocols, 25–31, 49, 56, 73, 76, 100; reassembly ethic of, 30, 66; rhetoric of domination, 34; roadshow for, 28, 65, 120; small group format in, 26–27, 26, 46–48, 56–64, 57, 81–82, 85, 93, 115–16; transnational, 28; visual productions of, 28, 65, 73–74, 87; whiteness informing, 37–38, 40–41, 45

feminist self help, products of: corporeal integrity, 48; demystification of the body, 26; ethic of individual responsibility, 42, 45, 48, 56, 118–20, 161, 172, 173; informed consent, 118; reassembly of authority, 48; reassembly of clinical encounters, 28, 118, 120; self-determination, 118; self-sovereignty, 33–35, 42, 45, 48

feminist self help clinics: antiabortion movement, effect on, 64, 66; defined, 25; described, 25; founders, 40; iconography, 25, 115–16; inaugural, 27; Pap smears in, 115–16, 119–20; power distribution, 118; state intervention, 63–64; transnationalization of, 27

feminist self help movement: beginnings, 46–47; black feminists in the, 65–66; central aim of, 173; emergence of, 7–8; founders, 46; goals of, 2; ideology, 35

feminist technoscience studies, 30–31, 32

Feminist Women's Health Center in Los Angeles, 64

Feminist Women's Health Centers, 26, 100

fertility: control over, women's desire for, 165, 167–68; economization of, 3–4, 17–18, 53, 153–54, 164; eugenic targeting of, 18; militarization of, 15–17; research on, 164–65, 169. *See also* reproduction

Firestone, Shulamith, 5–6

foreign aid: imperialism justified through, 16; for population control, 16–17, 20, 54, 163–70, 166

Foucault, Michel, 11, 13, 21, 24, 36–37, 175, 183n3, 192n10
Franklin, Sarah, 19, 159
Free Space (Allen), 84

Gage, Suzanne, 77, 78, *78*, *88*, *94*, 161, 195n58
Gainesville Women's Health Clinic, 44
Galison, Peter, 69
Garveyism, 58
gender, 19, 23, 90
generation as term, 185n17
Germain, Adrienne, 135–36, 138, 142
Global Strike for Women, 141
Goldberg, David Theo, 123–24
Greene, Jeremy, 51–52
group, groups: human relations research on children and, 59; as term during Cold War, 59. *See also* small groups format
Guattari, Felix, 29
Guevera, Che, 55
Gurner, Rowena, 190n7
Guttmacher Institute, 53
gynecology: beginnings, 106; patient-figure of, 106–7, 110, 112, 118; standard of practice, 123; transforming practices of, 110, 113–16, 118

Hammer, Fannie Lou, 42
Haraway, Donna, 73–74, 98–99, 186n23
Harding, Sandra, 98
Hartsock, Nancy, 98
health: contested site of, 55–56; counter-politicization of, 55; as an economic enterprise, 51–52
health care: biomedicalization of, 52–53; colonialism and, 16; health status vs. race and, 45; ay-providers of, 63–64, 95; legality of, 52–53; moral economy of, 101; protocols, standardized, reconfig-uring of, 51–52; revaluating the effi-ciency of, 121–23; state and, 52; strati-fication in access to, 19, 52, 107, 122;

transforming practices of, 5; universal, 122, 123
healthfulness, 94, 97
health moralism, 118
health: rights to, 50–51, 52; status of, 45
Hegel, Georg Wilhelm, 185n17
Helms Amendment of the Foreign Assis-tance Act, 168, 169
Herman, Ellen, 62
Hirsch, Lolly, 172
HIV research, 170
Hoffman, Frederick, 106–7
How to Stay Out of the Gynecologist's Office (pamphlet), 86
human capital, 51, 62
humanistic psychology, 61
human needs, hierarchy of, 61
human papilloma virus (HPV), 134, 135; vaccine against, 147–48
human potential movement, 61
human relations research, 59–62
human relations techniques, 63
human sciences, 3
human universal, eligibility for, 9–10
hypertext, 39

identity politics, 10, 14, 39, 45, 52
immigration policy, U.S., 18, 130, 147
immodest witnessing: objectivity in, 74, 87–88, 98–101, 178; vaginal self-examination, 73–79, *77*, *78*, 87–90
India, 17
industrial capitalism, 92
informed consent, 118
intellectual property and labor, 92, 160
International Fertility Research Program (FHI), 170
International Fertility Research Program (IFRP), 169, 170–71
International Monetary Fund (IMF), 145
International Planned Parenthood Fed-eration (IPPF), 163, 168, 172
International Pregnancy Advisory Servas (IPAS), 169

International Women's Health Coalition (IWHC), 135–46, 148, 174
interwingularity, 39–40
Inuit women, 124

Jain, Sarah Lochlann, 44
James, Selma, 141
Jane project, 46, 157
Johnson, Lyndon, administration of, 17, 18, 55
Jordanova, Ludmilla, 8
Journey into Self (film), 61

Karman, Harvey, 155–57, 159, 163, 168
Katz, Cindi, 100, 186n23
Keating, Peter, 71
Kennedy, John F., administration of, 60
Klawiter, Maren, 204n3
knowledge, situated, 98–100
knowledge production: counter-topographies of, 100; experience in, 80, 82–83, 98–99; information-sharing in, 27, 82–84, 85, 93, 159; oppositional tactics of, 98–101; and race in biomedicine, 123–27; the self in, 70; small groups format and, 27; standpoint theory of, 98–100; the subject in, 75; vaginal self-examination and, 68, 78

labor: affective, 92; expert, 71, 134; in feminist clinics, 63–64; politics, 63; of science, 75–76; social reproduction as unwaged, 92
LA Self Help Clinic, 40, 93
Latin American and Caribbean Women's Health Network, 136
Laufe, Leonard, 168, 169
lay health care, legality of, 63–64, 95
Lee, Laura Kim, 59
Lewin, Kurt, 59–60
life: alterability of, 3–4; biopolitics of, 36–37; capitalization of, 19–20; differential valuing of, 15, 44, 130, 131; politicization of, 29, 32, 36

Linnaeus, Carl, 9, 185n17
LINUX, 160
literature, black feminist, 44
lives less worth living, 16, 44, 130, 131
living-being: biopolitical formations of, 44; biovalue, 97–98; features of, 8, 10; feminism's concern with, 10; politicization of, 28–29, 32, 36; technoscience and politics of, 2–3, 26
Lorde, Audre, 45, 132–33, 141
Los Angeles, 40, 54–56, 58–61
Los Angeles and Orange County Feminist Women's Health Centers, 63
Los Angeles County Hospital, 54, 55
Los Angeles Feminist Women's Health Center, 73–74
Los Angeles Free Clinic, 55

macroeconomics, 17
Maginnis, Patricia, 157, 190n7
Malthus, Thomas Robert, 11
management, psychologization of, 60–61, 62, 63
Mao Tse-Tung, 58
marketing strategies of desire, 91–93
Marx, Karl, 5, 76, 80; Marxism and, 8, 29, 33–34, 58
materialization of the body, 86–87
Mbembe, Achille, 44
mechanical objectivity, 70
Medicaid, 18–19, 51–52, 63, 116, 121–22, 131, 147
medical-industrial complex, 5, 51, 54, 110, 112, 121
medical profession, 54, 55, 112
medical protocol, defined, 192n11
Medicare, 51–52
medicine: industrialization of, 19; politicization of, 55; racialized political economy of, 107; transformation of, 71
Memmi, Albert, 58
menstrual cycle: active control over, 160; mapping phases of, 108, 109, 110
Menstrual Cycle Study, 95–97

Menstrual Extraction (ME): abortion vs., 158–59; alterity in, 160; biopolitics, 161, 163; counter-empire project, 171–72; entanglements, 152–53, 171–75; exchange economy of, 159–60; global imaginary, 171–72; locations of geographic displacement, 160; Menstrual Regulation vs., 151–52, 163, 173–74; movement for, 172–73; noncommodification of, 171; protocols, 158–59, 161; as self-governed choice, 161, 172, 173; sharing component of, 159, 171; small group format, 161, *162*; sociality fostered through, 161; term usage, 151

Menstrual Extraction (ME) device: components, 150, *151*, *158*; noncommodification of, 159–60; Rothman design, 157–58; transnational distribution, 150; uses for, 150

Menstrual Regulation (MR): clinical trials, 169; entanglements, 152–53; feminist appropriation of, 175; Menstrual Extraction vs., 151–52, 163, 173–74; NGOs and, 168–69; research infrastructure, 169; statistics, 169

Menstrual Regulation (MR) device: Karman design, 157, 159; kits for, *167*; manufacture and funding, 169; patents, 156; transnational distribution, 153; use for, 151

Merck, 147–48

Mexican Americans, 124

micropolitics, 36

microscopy, 94–95

midwives, feminist, 35

Mitchell, Timothy, 17

modernity, 3–4, 18, 59, 154

modest witness, 74–75

Monthly Extract (newsletter), 172–73

Moraga, Cherríe, 45

Morales, Sylvia, 96

Morgan, Kathryn, 110

Morgan, Robin, 34

Mother's Anonymous, 44, 55

Mt. Vernon Group, 44–45

Myers, Natasha, 72

National Black Women's Health Project (NBWHP), 65–66, 126

National Cancer Institute, 110

National Institutes of Health, 23

National Training Laboratory for Group Development (NTL), 60

necropolitics, 13, 15, 16, 43, 107, 130, 131

Nelson, Ted, 39

neoliberalism, 21, 37, 130, 154

New Left ideology, 58

A New View of a Woman's Body, 96

New York Women's Hospital, 106

NGOs: empowerment, term usage by, 138; funding, 23; growth of, 18; Menstrual Regulation and, 168–69; NGO-ization of feminism and, 6, 31, 66, 105, 136–37, 146, 174; protocol feminism in, 31; transnational family-planning, 16, 20, 164, 168–70, 173

Nixon, Richard, 18

nuclear family, 16

objectivity: experience's role in, 80–83; historicizing, 69–70; of the immodest witness, 74, 87–88, 178; modes of, 70–71; reassembly of, vaginal self-exam and, 68–69, 99–100; in science and technology studies, 69

Office of Population (USAID), 163–65, *166*, 168

Office of Strategic Services, 59

operability, ethic of, 35

oral contraceptives, use statistics of, 113

oral health practices, 86

orgasm, 79

origami, biopolitics as, 186n23, 187n25

Our Bodies, Our Selves (Boston Women's Health Book Collective), 38, 84, 89

Pakistan, 17

Papanicolaou, George, 108–12

Papanicolaou, Mary, 108, 115

Pap smear: commodification of, 147; dual biopolitical logic, 121; entanglements, 102, 105, 114; feminist interventions, 104–5, 115–16, 127–28; historically, 105, 106; interpretation of results, 108, 116; introduction, 102–6; ontological politics of, 149; politics of, 115; rearranging gynecology, 106, 110; routinization of, 112, 119, 120–21; scale rearranging, 103–6; statistics, 113; subject-figures of, 114; technique of, 9, 108–10, 117

Pap smear screening: commodification of, 121; critiques of, 122, 125–26; demographics, 122–23; effectiveness, 102, 106, 121, 122, 136, 146; efficiency, reevaluating the, 121–23; frequency, recommended, 110, 119, 120, 123, 124–25; funding, 121; individualizing responsibility, 112–13; infrastructure requirements, 110, 112, 134–35; race and risk in, 103, 107, 122–25, 131–32; sexualizing incarceration, 131–32; socioeconomic stratification of, 131; spread of, 120–21; the state and, 121; transnational, inequity in, 141

participation, meaning of, 23

Pascoe, Peggy, 127

patient: as advocate, 71, 101; classed and raced, 130–31; clinical encounters, 28, 118, 120; moral economy of, 118; as observer/expert, 47, 97, 116; at-risk, norm of, 51–52, 112–13, 119–20; value of, 123. *See also* vaginal self-examination

patriarchy, 84, 92

"personal is political," 83–84

Petryna, Adriana, 20

pharmaceutical industry, 20, 51–52, 54, 71, 147–48

Phelan, Lana Clarke, 190n7

physicians: professional challenges, 71; women, percentage of, 129

Planned Parenthood, 18, 53, 54

political economy: biomedicine as a, 133–34, 149; problem-space of cervical cancer, 125–27, 133–34

politics, psychologization of, 58, 62

politics: of governed, 20–21; of politics, 36–37

poor, the: coercive management of, 4; economization of fertility, 17–19, 53–54, 138; health care, access to, 19, 51–52, 106–7, 114, 128. *See also* Medicaid

population, populations, 14; biocapitalization of, 20; bomb of, 16; biopolitics of, 13, 36; science, 17

population control: Cold War biopolitics, 15–17, 153, 154; economics of, 3–4; foreign aid funding, 16–17, 20, 54, 163–70, 166, 174

Population Council, 54, 146, 174

postcolonialism, 53

postscarcity activism, 41

Potts, Malcolm, 163, 168, 169

pregnancy tests, 27

prisons, Pap smear programs in, 131–32

Program for Action for a New International Economic Order (UN), 17

protocol, protocols: abortion, 28, 190n6; biomedicine, 29; consciousness raising, 26, 88; feminist self help, 25–31, 49, 56, 73, 76; Menstrual Extraction, 158–59, 161; standardized health care, 52–53; term usage, 29, 30; vaginal self-examination, 28, 68, 73–74, 191n8

protocol feminism: biomedicalization and the politics of health, 50–56; biopolitics and technoscience, 33–37; concerns of, 28–29; conclusion, 65–67, 178; emergence of, 31, 178; expression and spread of, 28; feminist self help as a form of, 28–31; infrastructure, 28; mapping clinical practices, 115–16; present day, 29, 30–31; racial politics, 37–50; reassembly ethic of, 65–66; social technology of small groups, 56–64; taking power through, 49; trend of radical

protocol feminism (*continued*)
experimentation, transformation of, 63–64, 66
psychographics, 91
public health care: biopolitics, 112, 130; neoliberal, elements shaping, 130–31, 175; politics of efficiency, 121–23

Rabinow, Paul, 57
race: biomedicine and, 124, 128, 130–33; indeterminacy of, 51; necropolitical work of, 44, 107; politicization of, 38; problem-space of cervical cancer, 103, 105–7, 123–28, 131–32
racial governmentality, 124, 127–28
racial state, formation of, 123–24
racism-sexism relation, police and, 43
radical as term, 191n9
radical feminism: base-superstructure model, 84; changes to, 174; feminist self help and, 33; knowledge-making practices, 76; Marxist influence on, 58; membership cohort, 114; the personal is political (slogan), 83; radical feminist health collectives and, 115; repoliticizing abortion, 155–58; term usage, 191n9
radical feminist consciousness-raising advocates, 58, 61, 92
Ravenholt, Reimert, 163, 165–67, 169, 171
Reagan, Ronald, administration of, 28, 37, 100, 122, 131, 133
recursive origami, biopolitical topology as, 186n23, 187n25
Red Thread, 140–41
regime of truth, 51
regulatory objectivity, 71
reproduction: alterability of, 1–4; capitalism and, 11; Cold War biopolitics regulating, 15–17, 153, 154, 163; economics of, 53; embodied, abolishment of, 5–6; entanglements, 7; multidimensional elements of, 6–7; necropolitics situating, 15, 44–45; politics of, 7, 9; seizing

the means of, 1–2, 4–5, 7–8, 35, 170, 177–81; term usage, 8–9, 185n17
reproduction, control of: ethic of individual responsibility, 2, 56, 160–61; in protocol feminism, 34. *See also* Menstrual Extraction
reproduction-production relations, 7, 8, 11
reproductive cycle, mapping of, 108, 109
reproductive health, 33; as a capital formation, 143–46; empowerment and, 138–44; IWHC policy, 139–40, 142–44; legislating, 147; politics of, 105; reassembly of, 49; RTIs in reconceptualizations of, 136, 138–43, 146; services of, 28; as term, 8, 31, 102, 105, 185n16; transnational models of feminist, 174
reproductive medicine, 97
reproductive sciences, 71–72, 108
reproductive techniques, 175
reproductive tract infections (RTIs), 136, 138–43, 146
research: clitoral study, 79; feminist self help advanced groups, 79, 86, 93; fertility, 164–65, 169; HIV, 170; human relations, 59–62; infrastructure, Menstrual Regulation, 169; outsourcing/biopolitics, 147; transnational economy of, 20
resistance, 183n3
Roberts, Dorothy, 44
Rodney, Walter, 140
Rodriguez-Trias, Helen, 45
Rogers, Carl, 61
Rossinow, Doug, 41
Rothman, Lorraine, 46, 157–58, 160, 171, 190n6

Sandoval, Chela, 98
Sanger, Margaret, 8, 33–34, 214n4
Santa Cruz Women's Health Collective, 34
Sarachild, Kathie, 201n24
Schabas, Margaret, 9

Schiebinger, Londa, 9

science, scientists, 70, 74, 75; demystification of, 75–76; moral economies of, 72, 79; transformation of, 71

science and technology studies, 5, 67, 69, 154

Seizing Our Bodies (Dreifus), 34

self-actualization, 61–62

self help: historically, U.S., 195n67; neoliberal formations of, 66

self help clinic, term usage, 26

"Self Help in a Suitcase" (road show), 80–81

sensitivity training, 61, 63, 65

service industry, 92

sex: alterability of, 2–4; politicization of, 36; spaces of transformation, 1

sex/gender system, 90

sexuality, vaginal self-exam and, 79

sexual violence, 168

Siemens, Albert, 171

"The Sisters Reply" (Mt. Vernon Group), 44–45

situated knowledge, 98–100

"Six Murdered Girls: Why Did They Die" (Combahee River Collective), 43

small groups format: clitoral research study, 79; consciousness raising and, 57; democracy formation and, 59; human relations research, 59–62; knowledge production in, 27; Menstrual Extraction, 161, *162*; self-collecting the Pap smear, 115–16; the shared as product of, 27, 82–84, 85, 93, 159; sociality fostered through, 161; social technology of, 56–64; vaginal self-examination and the, 26–27, *26*, 46–48, 57, 81–82

Smith, Andrea, 44

Smith, Barbara, 45

Smith, Beverly, 39, 132

Smith, Dorothy, 98

social change, 70–71

social collectivities, 57

social reproduction as unwaged labor, 92

social technology of group, 56–64

software development, 160

Somerville Women's Health Clinic, 38, 45

South African Black Consciousness movement, 58

Spivak, Gayatri, 23

standpoint theory, 98–99

Stanford Research Institute (SRI), 91

State Department, U.S., 60

Stephen, Marlene, *43*

sterilization, 18, 19, 54–55, 130, 163

Stoler, Ann Laura, 51

stratified biomedicalization, 52

structural objectivity, 70

Student Non-Violent Coordinating Committee (SNCC), *43*, 58

subject-figures: authentically normal, 41; ethical, feminist as, 9; experimental, 4, 48, 71, 170; the expert, 75; in identity politics, 14; the modest witness, 74–75; of the Pap smear, 114; of population control, 15–16; the scientist, 71, 74, 75; self-determining, 38; universalized, 7. *See also* patient

subjectivity, 70, 74

Summers, Lawrence, 17–18

sustainable farming, 5

technopolitics, 36–37

technoscience: affective economies of, 72–73; creation of, 71; entanglements, 2, 4–5, 7, 36; objectives and objects of, 71; politicization of, 67; reassembly of, 4, 7, 10, 21; situated, feminist topologies of, 98–101; vaginal self-examination as a form of, 69

technoscience-feminisms relationship, 2, 31

T Groups, 60

Third World Women Alliance, *43*

Thomson, Charis, 19

Timmerman, Stefan, 192n11

To Help Everyone (THE) Clinic for
Women, 55
topology, 11–12
transnational NGO, 16
Traweek, Sharon, 74

United Farm Workers, 55
United Nations, Program for Action for
a New International Economic Order,
17
UNIX, 160
unraced feminisms, 41–43
USAID, 15–16, 20, 163–70, *166*, 174
uterine cancer, 106

vaginal cells, classification system, 108–9
vaginal ecology, 94–95, *94*, 97, 119
vaginal self-examination: advanced
groups, 79, 86; aesthetic sensibility,
85; affective economies of, 73, 79–90;
affective entanglements of, 72–73, 79,
89–93; consciousness-raising power of,
47, 80–82, 100–101; counter-conduct
form of, 72; daily practice of, 93–96;
decline in, 100; facilitator's role, 73;
founding narrative of, 46–48; frequency, recommended, 120; the hand's
role in, 77–79, *78*; images of, *27*, 57, 81,
87, *88*; immodest witnessing, 73–79,
77, *78*, 87–90; the not uncommon in,
85–90; objectivity in, 68–69, 73; protocols, 28, 68, 73–74, 191n8; visual productions of, 73–74, 96
vaginal self-examination, products of:
demystification of the body, 76, 86;
healthfulness rearticulated, 94; knowledge production, 68, 78; materialization of the body, 86–87, 96–97; normal, reconfiguration of, 85–90, 96–97;
ontological collectivity of data, 87, 89;
rearticulation of healthfulness, 94,
97; reconfiguring the gynecological
exam, 115; refashioning objectivity, 68,
99–100; valuation of variation, 85–88

vaginal self-examination, small groups
format: founding narrative of, 46–48;
images of, *26*, 57, *88*; the shared as
product of, 82, 85, 93; web of mutual
summoning, 88
Values and Lifestyles Surveys, 91
venture technoscience, 92
violence: antiabortion movement, 64, 66;
racial, 42–43; against women, 168
"Virtual Speculums for a New World
Order" (Haraway), 99

waged labor, 92
Waldby, Catherine, 19, 97–98
Walton Report, 123
War on Poverty, 18, 55
Wasserheit, Judith, 136
Watts riot, causes of, 55
welfare recipients, 44, 55
welfare reform, 18–19
West Coast Federation of Feminist
Women's Health Centers, 65
West Coast feminist self help, 34
Western Behavioral Science Institute, 61
Western Training Laboratory (UCLA), 61
Whatley, Marianne, 129
When Birth Control Fails (pamphlet), 171
white, 1970s meaning of, 40–41
white feminist self help, 54
whiteness: cancer and, 106–7; feminist
self help movement and, 37–38, 40–41,
45; the necropolitical work of race and,
44, 107
"Why Did They Die?" (Combahee River
Collective), *43*
Wille-Muller, Stefan, 9
Wilson, Colleen, 199n125
Woman and Development Unit (WAND),
University of West Indies, 139–42
Women in Government, 147
women of color, 54, 55; as feminists, 42
Women's Choice Clinic, 160–61
Women's Community Health Center in
Cambridge, 63–64

Women's Community Health Center of Cambridge (WCHC), 39–40

women's health, 10, 29, 35, 39; centers of, 27–28, 63–64, 129–30

women's health movement: antiabortion movement, effect on, 89; cooptation, accusations of, 129; critiques of, 8; entanglements, 19; imagined community of, 183n2; legacy, 137; projects of, 21, 28; rearranging technoscience, 4; as term, 183n2

women's health services, funding, 174

women's studies, 14

Wonder Woman, 49, 50

Worcester, Nancy, 129

workforce: lay health care, legality of, 63–64, 95; racializations and genderings of, 62. *See also* labor

Working People's Alliance, 140

World Bank (WB), 145

World Fertility Survey, 184n5

World Health Organization (WHO), 23, 51, 112, 146

World War II, post–: birthrate regulation, 15; postscarcity activism, 41

Wright, Melissa, 44

Michelle Murphy is associate professor of history and women and gender studies at the University of Toronto. She is the author of *Sick Building Syndrome and the Politics of Uncertainty: Environmental Politics, Technoscience, and Women Workers*, also published by Duke University Press.

Library of Congress Cataloging-in-Publication Data
Murphy, Michelle (Claudette Michelle)
Seizing the means of reproduction : entanglements of feminism, health, and technoscience / Michelle Murphy.
p. cm. — (Experimental futures)
Includes bibliographical references and index.
ISBN 978-0-8223-5331-7 (cloth : alk. paper)
ISBN 978-0-8223-5336-2 (pbk. : alk. paper)
1. Women's health services — Political aspects — United States — History — 20th century. 2. Women — Health and hygiene — United States — History — 20th century. 3. Reproductive rights — United States — History — 20th century. 4. Women's rights — United States — History — 20th century. 5. Feminism — United States — History — 20th century. I. Title. II. Series: Experimental futures.
RA564.85.M88 2012
362.198100973 — dc23
2012011644